DOOR *to* DOOR

Also by EDWARD HUMES

Garbology: Our Dirty Love Affair with Trash

A Man and His Mountain: The Everyman Who Created Kendall-Jackson and Became America's Greatest Wine Entrepreneur

Force of Nature: The Unlikely Story of Wal-Mart's Green Revolution

Eco Barons: The Dreamers, Schemers, and Millionaires Who Are Saving Our Planet

Monkey Girl: Evolution, Education, Religion, and the Battle for America's Soul

Over Here: How the G.I. Bill Transformed the American Dream

School of Dreams: Making the Grade at a Top American High School

Baby ER: The Heroic Doctors and Nurses Who Perform Medicine's Tiniest Miracles

Mean Justice: A Town's Terror, a Prosecutor's Power, a Betrayal of Innocence

No Matter How Loud I Shout: A Year in the Life of Juvenile Court

Mississippi Mud: A True Story from a Corner of the Deep South

Murderer with a Badge: The Secret Life of a Rogue Cop

Buried Secrets: A True Story of Serial Murder

DOOR *to* DOOR

The MAGNIFICENT, MADDENING, MYSTERIOUS WORLD *of* TRANSPORTATION

EDWARD HUMES

HARPER

An Imprint of HarperCollins*Publishers*

 For information, address HarperCollins Publishers, 195 Broadway, New York, NY 10007.

HarperCollins books may be purchased for educational, business, or sales promotional use. For information, please e-mail the Special Markets Department at SPsales@harpercollins.com.

FIRST EDITION

Designed by William Ruoto
Frontispiece courtesy of Mmaxer/Shutterstock

Library of Congress Cataloging-in-Publication Data has been applied for.

ISBN: 978-0-06-237207-9

16 17 18 19 20 OV/RRD 10 9 8 7 6 5 4 3 2 1

This book is dedicated to my favorite door in the world—the one that leads to my family.

Contents

DOOR *to* DOOR

Introduction

THE THREE-MILLION-MILE COMMUTE

The traffic apocalypse arrived at last on a humid and smoggy Friday evening in Los Angeles in 2011.

Carmageddon, we called it. Even Metro LA, the relentlessly optimistic transportation agency charged with keeping this capital of car culture moving, finally gave in and adopted the end-of-days nickname for the traffic disaster it had planned and scheduled.

Carmageddon was invented to bring relief to the most traffic-jammed highway in America. Instead, it sparked an unexpected revolution.

Long dreaded by Los Angeles commuters, truckers, cops, and business owners, Carmageddon at least began according to plan. Shortly after rush hour's heavy traffic had dwindled into Friday dinner-hour light, a swarm of orange-vested highway crews assembled amid a chorus of slamming pickup truck doors. They wrestled blockades into place at the entry ramps onto Los Angeles's busiest freeway exactly as scheduled. Their battle plan had been designed by a military engineer who rebuilt bomb-cratered highways in Iraq before coming to Metro, and his timetable was as mercilessly exacting as a Marine Corps march. By midnight, a ten-mile, ten-lane stretch of Interstate 405, which carries nearly 400,000 cars and trucks a day, had become a vehicle-free zone for the first time since the governor of California cut the ceremonial ribbon in 1962.

The next morning, hundreds of Angelenos gathered on overpasses up and down the freeway to gawk at what had been for generations a ceaseless, roaring river of cars, now transformed into a silent and empty ribbon of asphalt curling gracefully over the Sepulveda Pass, vanishing into the summertime haze. You could hear birds singing that morning where the singing of tires on asphalt had so long reigned.

The goal was simple: to fix one of America's worst commutes.[1]

The project would graft a new lane onto the overburdened freeway to relieve chronic congestion and crashes, the result of jamming twice as many cars and trucks on the highway at peak hours as its designers envisioned. The closure was to let the bulldozers rule the road for an entire weekend while portions of three bridges spanning the freeway fell to the wrecking ball. It also meant a half million weekend drivers would just have to find another way. Which was, of course, why the event had been dubbed Carmageddon: those drivers had no other way to go. Or so it seemed.

The 405 has long been an object of dread in the region, at once essential, unavoidable, and reviled. In other parts of the city, parallel freeway routes give freight haulers and commuters multiple options as they traverse the LA sprawl. But the 405 is the only direct major route linking West Los Angeles, UCLA, and the coastal communities to the south with the San Fernando Valley and its suburban commuters to the north, as well as the resort towns of Ventura and Santa Barbara farther up the coast. Having only one way to get between major population, recreation, and employment centers in a city with 6.5 million cars—and a historical disdain for both public transit and carpooling—has long been a recipe for epic gridlock.[2] The standing joke is that its numerical name stands for the "four or five miles per hour" speeds the 405 inflicts on drivers. Regular travelers on the 405 know, and meters and measurements have confirmed, that this is really no joke at all but the actual speed of traffic at peak hours and places on this less-than-"free" way. Tesla Motors founder Elon Musk branded his commute through 405 traffic "soul killing," inspiring him to vow to reinvent the commute and to make public his *Jetsons*-style "hyperloop" concept: LA to San Francisco in one half-hour pneumatic whoosh.

The entire expansion would take five years and $1.3 billion to complete before the promised traffic relief arrived, with the diciest part being those initial fifty-three hours when the freeway would be emptied completely. To get the gain, the city fathers warned, there had to be pain: the shutdown was universally predicted to turn the rest of LA into one big parking lot. All those cars, all those big rigs, all those people and goods normally occupying the 405, would flood other routes ill-equipped to absorb the load. Paralyzing jams, angry and reckless drivers, more accidents, smog, and deaths were predicted as the city and region prepared for the worst.

Hospitals beefed up their emergency room staffs and set up beds in disused wards to handle the overflow. The mayor begged drivers to park it for the weekend, to leave the car keys on the table. Thirty Hollywood celebrities (think Kardashians, Captain Kirk, and the host of *American Idol*) were recruited by the Los Angeles Police Department to send apocalyptic warnings to their combined 100 million Twitter followers. In a City Hall bunker called ATSAC (Automated Traffic Surveillance and Control) the transportation overlords tweaked 4,532 stoplights to shunt cars away from the 405. The same computer cowboys are Oscar Night's secret weapon, slowing "ordinary" drivers so lines of Hollywood A-lister limos arrive on time. War rooms at FedEx, UPS, and the Port of Los Angeles's sprawling truck terminals rerouted everything they could. But even these masters of the transit universe knew they could not alter the simple physics of too many cars in too little space: they all predicted the event would be the worst man-made disaster in LA history. Carmageddon would earn its name.

Except it didn't. Not one dire prediction came true.

Congestion *decreased* that weekend citywide. There were no major traffic jams. No spikes in crashes or deaths. No escalating road rage or extra ER patients.

Magicians in a bunker didn't do this. Car-centric Angelenos did. They walked, biked, rode Metro Rail, took the bus, ride-shared with Uber, Sidecar, and Lyft, and found their formerly "essential" drives to be quite optional. Smog in the 405 corridor dropped to a tenth of normal and the entire city inhaled 25 percent fewer air pollutants—for an entire week as the event's halo effect persisted. The day-after buzz focused not on traffic jams and frustrated travel plans but on how the city might find a way to make the Carmageddon effect permanent. "Carmaheaven," headlines renamed it.

To say many in the city were stunned, from the professional traffic czars to everyday commuters, would be an understatement. The 405, like most of the interstate system, had been carved into the city's bedrock with the explicit purpose of appeasing and pleasing drivers, with no concessions or thought given to the barriers it would create for walking, biking, livability, community preservation, or health. Congestion was supposed to be fixed by *expanding* such car-centric public works—not by closing them down.

Yet this is exactly what happened on that steamy Carmageddon weekend of July 15, 2011. Trends that had been hinted at for years—how Americans, particularly the young, had begun forsaking car ownership and even driver's licenses in favor of mobility by any means handy, whether bike, rideshare, bus, or train—suddenly seemed to be playing out in real time on the streets of Los Angeles. Every traffic truism held dear for the past sixty years had been turned on its head, from the cures for gridlock to the most effective use of scarce transportation resources. When closing lanes lessens congestion instead of causing it, that means something. And when it happens in the second biggest city and the most important port city on the continent, the world notices. Soon LA leaders began a long and painful reconsideration of transportation policies, a struggle to replace car-centric planning with a focus on broader options for mobility. But the implications of Carmageddon went far beyond one city. They were national, global, and universal.

In Los Angeles of all places—the city perhaps most profoundly shaped by and dependent upon the automobile—America's love affair with the car was being openly questioned. The challenge came not from gadflies and outliers but from the mayor and councilmen and the chamber of commerce. With Carmageddon, the lifeblood of our economy and way of life, the movement of goods and people from door to door, had reached an unexpected tipping point.

More than smartphones, more than television, more than food, culture, or commerce, more even than Twitter or Facebook, transportation permeates our daily existence. In ways both glaringly obvious and deeply hidden, thousands, even millions of miles are embedded in everything we do and touch—not just every trip we take, but every click we make, every purchase, every meal, every sip of water and drop of gasoline. We are the door-to-door nation.

What does it take to keep America moving—to keep our cars and trucks on the road, our store shelves stocked, our cupboards and sock drawers, gas tanks and coffeepots full? Like an iceberg, impressive on the surface but with far more obscured below, the machinery of movement grinds away, a heady blend of miracles and madness that enables Americans to drive 344 million miles every hour[3] and move $55 billion worth of goods every day[4] (which, by way of comparison, is triple the daily income of every household in America combined). As big as Carmageddon was, it's just a sliver compared to the global movement machine that supports the way we live, work, and prosper—and that we can no longer live without, even for a handful of days. Look anywhere in the modern American home and you will see necessities, ordinary items—things we use every day—whose journeys to us make Marco Polo's voyages pale by comparison.

So what does this look like under the hood, this vast system that takes us and our stuff door to door day in and day out? What does it truly cost us, and what can we afford moving forward? Are the wheels spinning better than ever, or are they about to come off the rails? Or is it a bit of both?

These seemingly simple questions gave rise to this book. Think transportation detective story, except the goal is not unmasking a villain but shedding light on some of the hidden characters, locations, myths, and machinery driving our buy-it-now, same-day-delivery, traffic-packed world.

Consider the mostly hidden world of car crashes, where the physics of impact meet the reality of drivers drinking, texting, and speeding—where four Americans die every hour, and one is injured every 12.6 *seconds*. One of the many hidden parts of this violent side of the door-to-door world: we know how to make driving far safer, and yet we do not do it. A car in theory is just a shipping container for people, which should make it an easy thing to engineer and deploy for maximum safety and efficiency, like a FedEx mailer or stackable cargo containers—foolproof, protective, and boring. But cars have evolved into objects of culture, power, status, desire, image, and habit. Do you know why the standard width of a modern car is six feet? Brilliant engineering? Efficient use of materials? No, it's simply habit and tradition, based on the now-irrelevant fact that six feet is the minimum width needed for a carriage drawn by two horses hitched side by side. It should surprise no one that such a mélange of factors yields vehicles in which engineering and safety have little primacy and, as the saying goes, mileage may vary. Life expectancy, too.

Then there's the hidden world of Angels Gate, the hilltop vantage point in Los Angeles overlooking North America's two leading, constantly bustling side-by-side seaports awash with cargo aboard ships the size of city blocks. There you can find a single mother of four by the name of Debbie Chavez, author of the daily "Master Queuing List," with which she quite literally holds the lion's share of the U.S. consumer economy in her hands. Angels Gate is where America's commute truly begins, aboard ships piled fifteen stories high with cargo containers, each far safer than your car and each hauling enough products to stock five Costco warehouse stores. On any given day, fifty such vessels are lingering at the dual ports, with a couple dozen more waiting in line for space at the docks. Waiting for a spot on Debbie Chavez's list.

Then there are the steely-nerved port pilots who live (and risk death) by that list as they guide the towering ships past Angels Gate and up to the biggest docks in America. With the skill of surgeons they squeeze these mega-ships beneath bridges with just a few feet of clearance and down narrow channels intended for the smaller vessels of a bygone age—an exercise in titanic parallel parking not for the fainthearted. Wonder and madness, the new and the old, are constantly battling in the door-to-door maelstrom.

Consider the tiny country that many of these towering ships call home, host to the most powerful oceangoing fleet in the world. This vast corporate flotilla that rules the seas without a single gun, cannon, or missile. All that fleet does is carry 15 percent of the world's goods. And through its alliance with the number two shipping line, it controls a third of the world's products. That's the hidden universe of door-to-door power.

Finally, there is the sheer crazy wonder of it all. The capacity to transport a supercomputer, a desperately needed medicine, or a tube of toothpaste from a factory in Shanghai to a store in Southern California or New Jersey or Duluth—and to do so 20 billion times a day reliably, affordably, quickly, and trackably—may well be humanity's most towering achievement. And yet it is accomplished with so little fanfare it's virtually hidden from public view.

Every time you visit the Web site for UPS or Amazon or Apple and instantly learn where in the world your product or package can be found and when it will thump on your doorstep, you have achieved something that all but the still-living generations of humanity would have declared impossible or demonic.

Consider the travels of that humblest yet most vital elixir: the morning cup of coffee. The day I started working on *Door to Door,* I began my research in the nerve center, fuel depot, energy source, and morning assembly zone Americans know best: the kitchen. In the cupboard I found a half-used bag of Starbucks-roasted, Costco-sold French Roast coffee beans. This is known in the trade as a "private label" product: coffee made by a major brand company, in this case, Starbucks, but for labeling and sale by a grocery or warehouse chain under its own name and private brand sold nowhere else. America's second favorite drink (not counting water) seemed the perfect place to start looking for the transportation "footprint" of stuff.

Costco French Roast consists of a blend of beans from South America, Africa, and Asia, each component shipped by container vessel up to 11,000 miles in 132-pound loosely woven sacks of raw, green coffee beans, some across the Pacific Ocean to ports up and down the West Coast, the rest via the Panama Canal, perhaps the Suez Canal, then on to one of several East Coast ports. The complexities are so great on this routing—based on ship space, season, and the vagaries of rates and departures—that it's difficult to trace bulk products more precisely than this. The raw beans then travel by freight train or truck (2,226 miles for the Port of Los Angeles portion) to one of the world's largest blending and roasting plants, located at 3000 Espresso Way in York, Pennsylvania, one of six such plants in the Starbucks empire and the one identified by the company as principally serving Costco. After roasting, blending, and testing to make sure every batch smells and tastes exactly the same no matter how many times a customer buys Costco French Roast, the beans are sealed in plastic and foil composite bags with their own coast-to-coast mileage footprint. Then the packages are stacked on wooden pallets (sourced from all over the nation) and shipped another 2,773 miles back across the country to the Costco depot in Tracy, California, from which my coffee was trucked to my local Costco store. By the time I got those beans, they had traveled more than 30,000 miles from field to exporter to port to factory to distribution center to store to my house—more than enough to circumnavigate the globe.

But that's not where the coffee mileage stops. There are the components of my German-built, globally sourced coffeemaker, which collectively traveled another 15,700 miles to reach my kitchen. My little bean grinder had a similar triptych. The drinking water I use to brew my coffee comes to my home from a blend of three sources: from groundwater pumped in from local wells about 50 miles distant; via the 242-mile Colorado River Aqueduct; and through the 444-mile California State Water Project, which moves water south from Northern California, forces it 2,000 feet straight up and over the Tehachapi Mountains, then down into Southern California. The fuel and energy required for this third leg exceeds the electricity demand of the entire city of Las Vegas and all its glittering casinos. The electricity that powers my coffee machine runs through a grid festooned with millions of transformers and capacitors, most of which are now imported across 12,000 miles from China through the ports of Los Angeles and Long Beach, a complex that is a veritable city unto itself. The natural gas that fuels the power plants that provide most of the electricity to my coffeemaker is obtained from gas fields in Canada and Texas and sometimes farther through a 44,000-mile network of underground pipelines—North America's hidden energy transport plumbing.

At this point the collective transportation footprint on my cup of coffee is hovering at 100,000 miles minimum. And that's not counting the seemingly smallest segment of that journey, my 6.3-mile drive to Costco in my 2009 Toyota Scion xB, which has the most massive transportation footprint of anything I own—and not because I drive it very far. I chose to buy a used vehicle on the theory that a secondhand but fuel-efficient conventional car is greener and less wasteful overall than a newly made hybrid or electric (not to mention a whole lot cheaper), and because we needed something big enough to hold our three greyhounds (which it does, barely, with two humans on board, too). The Scion was built in Japan out of about thirty thousand globally sourced components from throughout Asia and Europe, with one U.S. manufacturer contributing: the tires are from Ohio-based Goodyear, which has factories in Asia as well as the U.S. The assembled car was shipped from Japan to the Port of Long Beach in California, then trucked to a dealership in Southern California (other cars arriving by ship move by train to more distant dealers). The cumulative travels of the raw materials and parts of my car totaled at least 500,000 miles before its first test drive. The gas in its tank is a petroleum cocktail that adds another 100,000 miles to the calculation, as the California fuel mix consists of crude oil from fourteen foreign countries and four states.[5] Most of this oil arrives by tanker ships at West Coast ports, then moves thousands of miles around the state and country to tank farms, refineries, fuel depots, and distribution centers via pipelines, railroads, canals, and semitrucks before finally appearing at my neighborhood gas station.

Thousands of man-hours and billions of dollars in technology and infrastructure—along with the efforts of countless unsung heroes who pack, lift, load, drive, and track it all—combined to bring that cup of coffee to my lips (and my wife's nightstand; I'm the morning person in our household).

That cup of coffee is a modern miracle, magical and mundane at the same time, though we hardly if ever notice the immense door-to-door machine ticking away, making it happen with product after product, millions of them, each requiring the same level of effort and movement, day after day.

Elon Musk finds the average twenty-seven-mile commute on the 405 soul-killing, and it's hard to disagree. Yet our true daily commutes are so much more than that. Our true daily commutes, beginning first thing in the morning with the travels of my cup of coffee—and followed by my socks and orange juice and dog food and dish soap—are more on the order of 3 million miles.

We live like no other civilization in history, embedding ever greater amounts of miles within our goods and lives as a means of making everyday products and services seemingly *more* efficient and affordable. In the past, distance meant the opposite: added cost, added risk, added uncertainty. It's as if we are defying gravity.

The logistics involved in just one day of global goods movement dwarfs the Normandy invasion and the Apollo moon missions combined. The grand ballet in which we move ourselves and our stuff from door to door is equivalent to building the Great Pyramid, the Hoover Dam, and the Empire State Building all in a day. Every day. It is almost a misnomer to call this a transportation "system." Moving door to door requires a complex system built of many systems, separate and co-dependent, yet in competition with one another for resources and customers—an orchestra of sometimes harmonizing, sometimes clashing wheels, rails, roads, wings, pipelines, and sea lanes.

And yet the same conductors of this magical door-to-door symphony also leave us with grinding commutes: Los Angelenos spent an average of eighty hours a year stuck in traffic in 2014, while New Yorkers bore a "mere" seventy-four hours of traffic-jam penance.[6] They gift us with more than 35,500 annual traffic fatalities—one death every fifteen minutes.[7] And we are the proud owners of roads we can no longer afford to maintain, saddling the country with an impossible $3.6 trillion backlog in repairs and improvements to aging roads and bridges[8]—a deficit that grows every year, because Congress has refused to raise the 18.4-cent-per-gallon federal fuel tax.[9] It's been frozen in place since 1993 without even adjustments for inflation (which means that 18.4 cents is now worth only 11.2 cents). Drivers now pay barely half the cost of maintaining the highway system though gas taxes, with the rest subsidized. Meanwhile, more than 61,000 of our bridges are officially designated "structurally deficient," a third of which are deemed "fracture critical."[10] There is no money to fix most of them.

How can a country that deploys insanely capable robot rovers to Mars and puts unerring GPS chips in our pockets leave us with two-ton rolling metal boxes to transport one person to work each day—boxes that kill ninety-seven of us every day and injure another eight every *minute*? Cars are the American family's largest expense after dwellings, our least efficient use of energy, the number one cause of death for Americans under thirty-nine, and our least productive investment by far. The typical car sits idle twenty-two hours a day, for which privilege Americans, on average, pay $1,049 a month in fuel, ownership, and operating expenses.[11]

Miraculous and maddening—these are the two poles of our transportation system, inside which Americans spend an unimaginable 175 billion minutes a year, at an equally unimaginable annual cost in money and time of $5 trillion.[12] An estimated $124 billion of that cost consists of lost productivity due to traffic jams, projected to rise 50 percent by the year 2030.[13]

These two faces of transit are often viewed and treated as two separate, even competing worlds—the frequently frustrating, in-your-face reality of how we move ourselves, and the largely hidden world of goods movement with its gated marine terminals, secure distribution centers, and mile-long trains with unfamiliar foreign names on the container cars: Maersk and COSCO and YTL. The same Los Angeles–area communities that embraced a billion-dollar bill to add a lane to Interstate 405 have successfully fought off for fifty years the completion of another north-south freeway that would connect the port to inland

California with its vast web of warehouses, distribution centers, and shipping terminals. Residents oppose the building of the last five miles of this freeway, Interstate 710, because it is seen as benefiting freight, not people, as if the local Walmart stocked itself. The stream of big rigs flowing from the port instead have to take roundabout and inefficient routes on other freeways, wasting fuel and time—and adding to commuter traffic jams as well, where drivers curse the ponderous big trucks they have inflicted on themselves.

Perceptions aside, there are no separate transportation systems, just one, sharing the same crumbling, underfunded infrastructure, the same limitations of cost, energy, and environment. They are two limbs on a single tree, mutually dependent, both grappling with potential disruption from old problems and new technologies. The idea that goods and people can be considered separately from one another when it comes to building roads or enlarging ports or creating supply chains is one of our primary transportation obstacles and myths.

There are many other common myths that simultaneously drive and cripple our door-to-door machine: bigger, heavier cars are not safer. They make our roads *more* fatal. Mass transit isn't the most subsidized mobility option, as most Americans believe. Your car is. Except for highly trained professionals and the occasional savant, there are no good drivers—just bad and less bad ones. The very term "accident" is a lie we tell ourselves, as almost all crashes result from purposeful negligence, recklessness, or law breaking.

And the biggest myth of all, one that distorts how we plan and spend, is the one exposed by Carmageddon: adding more lanes does not, in the end, fix traffic.

Not only did Carmageddon do the opposite, improving congestion (at least temporarily) with the *closure* of lanes, but the completion of the five-year project did not achieve the desired outcome of less congestion, either.

One year after the additional lane on the 405 opened, commuters on that ten-mile stretch of freeway spent an average of one minute *longer* stuck in traffic than they did before the new lane was added.[14] Call it the *Field of Dreams* syndrome: if you build it, the cars will come. Traffic, like nature, abhors a vacuum, and the lure of the extra lane brought in more cars than ever. The 405 remains the commuter's bane, worse than ever—more than a billion spent, with nothing to show for it.

Carmageddon marked a tipping point nonetheless, part of a larger convergence of trends and ideas that are transforming the way we move door to door. For the first time in a century, the strategy of designing our built landscape around cars and their free flow is being challenged. One in ten homes is car free now.[15] The Millennial generation is shunning licenses and car-dependent suburbs in increasing numbers. Americans are driving less per capita. They're shopping outside the home less, too. Ride-sharing apps, crowdsourced transit guides, bike commuters, and Wi-Fi–enabled buses are taking on the likes of General Motors, Yellow Cab, Hertz, and the bureaucracies that protect them. Every car added to the ridesharing fleets of Uber and Lyft and the others leads to thirty-two fewer car sales[16]—a seismic force in the auto industry. And in an unmistakable sign of the times, car-friendly Los Angeles did the unthinkable, hiring a new head for the city's transportation department by stealing her from in-state rival San Francisco, where she had become famous (or infamous) for her dedication to pleasing pedestrians, bicyclists, and transit riders as much as drivers.

Then there's the wild card in the mix: the driverless cars (and trucks) now being developed and tested on highways and city streets. They appear to be just a few years away from consumers, with the potential to change everything—from drastically reducing highway deaths, to transforming car ownership as we know it now, to ending the need to devote a third of our urban space to parking. Think about that a minute: the most interesting thing about autonomous vehicles is not what they do while taking you somewhere. It's what they do *after* they drop you off. Carmakers are racing to bring these wheeled robots to market even as the technology terrifies them as an extinction threat to their century-old business model—and to the millions of jobs that business model creates. Every shiny new evolution in our door-to-door world has exacted a price along with its promise: the invention of the steamship that ushered in the first wave of globalization; the discovery that gasoline could be more than its first incarnation as a remedy for head lice; the containerization of cargo that enabled the second wave of globalization—our age of outsourcing and offshoring. The game changer of removing human drivers from the transportation equation could easily be the most disruptive evolution of all, fraught with as many challenges and fears as opportunities and boons.

The hidden side of our commute, the flow of goods, has become so huge that our ports, rails, and roads can no longer handle the load. They desperately need investments of public capital that the nation does not seem to have. Yet it's an investment that must be made, as logistics—the transport of goods—is now a vital pillar of the U.S. economy. Goods movement now provides a greater source of job growth than making the stuff being shipped.

"Your kids will never go hungry if they have degrees in global logistics," says the head of UPS for the American West. "But we have to leave them a transportation system that works."

At the same time, new manufacturing technologies—the science fiction turned fact that is 3-D printing—are pushing in the opposition direction. This "unicorn" technology gives businesses in Brooklyn, Boston, and Burbank the power to manufacture a fantastic range of products—from surgical implants to car parts to guns—and to do it cheaper than a Chinese factory can 12,000 miles away. New local businesses are emerging almost daily, stealing manufacturing from offshore factories and the eyes of Angels Gate. It's just a commercial trickle, this "re-shoring" of offshore manufacturing, not enough even to register statistically, much less hurt the business of ports and shippers—for now.

These countervailing trends have the power to upend our brilliant and terrible global flow of goods and people, not in some misty, speculative future, but in a very few years. Which means our transportation-immersed door-to-door world and every aspect of it—culture, food, economy, energy, environment, jobs, climate, your cup of coffee in the morning—is at a very large, very vital fork in the road.

That's the commute we're all riding now, and whether the choices and trends now in play lead to a global Carmageddon, or Carmaheaven, or a bit of both, is one of the great unknowns of our age.

As one of the nation's leading transportation scholars, authors, and bloggers, David Levinson of the University of Minnesota, sees it: "We've been slow to change. But change is coming."

Buckle up.

Chapter 1

MORNING ALARM

My day starts with a bong—the deep, insistent pull of Big Ben's bell toll, waking me courtesy of my smartphone. I haven't used my bedside alarm clock in years, yet another specialized gadget shunted aside by a jack-of-all-trades iPhone, the clock relegated to collecting dust on my nightstand and not much more.

And so my commute begins before I leave the house. Add 160,000 miles of travel to the tally of my daily journeys. That, at bare minimum, is what it takes to bring an iPhone from origin to customer.

My phone sounds a second alarm (the *Dive! Dive!* klaxon of a Navy submarine) when it's time to wake my son for school, then a follow-up alert to remind me to prod him toward the door in case he fell back to sleep. A fourth alarm informs me when it's time to bring that wake-up cup of coffee to my wife. The fifth and final preset alarm—I choose the iPhone's annoyingly screechy "Sci-Fi" tone for this one—tells me to give the cat his insulin shot (yes, cats get diabetes). This is not the favorite moment of the day for either of us, but family Morning Person gets veterinary medication duty by default, so both cat and I put up with the unwanted intimacy of the hypodermic pen.

In between the various alarms, the smartphone allows me

to peruse several morning newspapers, check e-mails, read news feeds and Twitter lists, attend to online banking or bill pay, and, as I go about my morning routine of coffee making and dog feeding, listen to either National Public Radio or my latest audiobook. Today is also my wedding anniversary: time to check with the florist's Web site to verify the bouquet I ordered remains on schedule for delivery at Donna's office later this morning. A little later the phone's Google Maps app will direct me to the right time and place to catch a bus to a conference in San Pedro on the future of ports. (Hint: that future is both promising *and* troubled.) After the conference, another handy app will summon a driver from the rideshare service Lyft for the return trip home. Serendipity being what it is, my driver will turn out to be a moonlighting longshoreman with an entirely different view on the future of ports. He sees them as rapidly changing, frequently unsafe, roiled by congestion and labor disputes, but, most of all, as a haven for a vanishing breed in America: a truly financially rewarding path for a blue-collar middle class. His smartphone is his vital link, too, for working in both harbor and rideshare universes.

Such high-tech wonders as the iPhone have extraordinarily complex and far-flung supply chains. The specifics of suppliers and components are often closely held company secrets, but starting in 2012, Apple began publicly disclosing information on its top two hundred suppliers.[1] What that in turn reveals: smartphones may have more transportation embedded in their production and distribution than almost any other widely used consumer product other than the modern car, which has more (and far larger) globally sourced parts in its recipe.

Unlike cars, smartphones are also prodigious transportation *reducers*. They have accelerated the substitution of digital newspapers for physical newsprint, which otherwise would have to be transported by rail and semitruck to the presses, then physically delivered to homes and points of purchase after printing. The energy and carbon footprint of a single copy of a newspaper is roughly the same as driving a car one kilometer[2]—not much on its own, but it adds up pretty quickly over time and across whole subscriber bases. Phones do the same for the physical media of books and music as well as for paper bills and the physical transportation of the proverbial checks in the mail. In the U.S., about half of smartphone owners bank and pay bills with their phones,[3] which amounts to quite a few people going paper-free (and therefore transportation-free) on payday and balance-due day. Of all 243 million adults in the U.S.,[4] 64 percent own smartphones. For adults under fifty, the smartphone adoption rate is a stunning eight out of ten Americans.[5] Put the numbers together and it turns out that 78 million American adults are going paper-free and travel-free when it comes to bills and banks, at least some of the time. That's a big change from essentially zero in 2007.

Beyond that, smartphones have become the Swiss Army knives of the tech world, displacing a host of specialized devices: music players, alarm clocks, radios, cameras, calculators, tape recorders, GPS navigation devices, calendars, date books, Rolodexes, handheld gaming devices, metronomes, egg timers, flashlights. When smartphones and apps supplant a stand-alone device, that's not just one less thing we have to carry around personally. It's one less thing someone else has to carry to us. In short, thanks largely to smartphones, the bottom has fallen out of whole categories of products, business models, and jobs—and their corresponding transportation footprints.

And then there are the travel and mapping smartphone apps that simplify, and thereby encourage, the use of mass transit, ridesharing, biking, and even walking. They don't shrink transportation per se, but such technology can shift drivers from cars to less wasteful, less energy-intensive options that, among other things, lower demand for parking. This is no small feat. Cities that have studied urban traffic in recent years have found anywhere from 30 to more than 60 percent of drivers on congested city streets are clogging things up because they are searching for parking.[6] This is one of the many counterintuitive aspects of moving door to door: the process of taking your car out of traffic can make traffic worse. This rule of the road flips, however, if an app points drivers efficiently and quickly to waiting parking places without all the searching, jockeying, and frustrated braking when an enticing spot turns out to have red paint on the curb. Other apps guide drivers to the least congested routes, which doesn't shrink miles—it may even add them—but it does shrink time on the road and keeps drivers from making traffic jams worse, thereby easing congestion and gasoline consumption for all.

On the other side of the ledger, the birth of smartphones is an exercise in profligate transportation consumption. Their twenty-first-century materials, along with the routine, even blasé sourcing over vast distances, overturn the methods and economies of the past, which generally shunned distance as an added cost and risk. Not so the iPhone. Its components collectively travel enough miles to circumnavigate the planet at least eight times before the phone receives its first call or sends its inaugural text.

Cracking open my iPhone 6 Plus—Apple's version of the big-screen "phablet"—reveals not just a marvel of globally sourced miniaturization but also a high-tech road map that touches just about everywhere. Along with the processor and graphics chipset and the rechargeable battery (the most massive internal part), there is a long list of individually sourced components: two cameras, a video recorder, a digital compass, a satellite-navigation system, a barometer, a fingerprint scanner, a high-resolution color display, an LED flashlight, touch sensors, a stereo system, a motion sensor/game controller, encryption circuits, an array of radio transmitters that connect via Wi-Fi, Bluetooth and near-field communication bands, and, last and also least, the guts of a cellular telephone.

At least two dozen primary suppliers on three continents and two islands (Japan and Taiwan) provide these parts.

The transportation complexity is magnified further because many components do not move in a simple path from supplier to final assembly. Some go on a hopscotching world tour from one country to the next and back again as one piece is joined to another to create an assembly, which is then moved elsewhere in the world for another part to be inserted or attached. The phone's innards are put together much as a cook assembles ingredients for a dish that becomes, in turn, a component of another chef's course, which is then incorporated by someone else into a larger meal. Ingredients move back and forth from high-tech equivalents of refrigerator, cutting board, stove, and plate.

The fingerprint sensor embedded in the iPhone's home button—Apple's Touch ID system, which allows a fingerprint scan to replace a typed password—is a good example of this sort of *Top Chef* supply itinerary.

The home button journey begins in Hunan province, China, at a company called Lens Technology, Ltd., in the city of Changsha, where superhard transparent artificial sapphire crystal is fashioned into the button cover. This is the part of the button an iPhone user physically touches, made of the same synthetic sapphire used in high-end watches, avionics displays, and missile systems because of its near–diamond-like hardness, durability, and scratch resistance. The sapphire cover is then bonded to a metal trim ring brought 550 miles from the LY Technology factory in Jiangsu province, and then shipped 1,000 miles to the Dutch-owned NXP Semiconductors assembly and testing plant in Kaohsiung, Taiwan. There the sapphire-metal ring combo is married to a driver chip imported from a Shanghai factory (another 600 miles) and a Touch ID sensor chip from an NXP silicon wafer fabrication plant in Europe, which tacks 5,000 more miles onto the itinerary.

Next, a button switch imported from a Panasonic subsidiary is brought in 1,500 miles from Japan, along with the springlike plastic component called a "stiffener" from a Shanghai factory (another 600-plus miles) owned by the American company Molex. These pieces are combined at another Taiwanese manufacturer, Mektec, which adds in its own part, called the flex circuit.

Mektec then ships this assembly 1,500 miles back to Japan, where a plant run by technology giant Sharp laser-welds all the pieces into a sealed and functional Touch ID module. The completed assembly ships about 1,300 miles to the Foxconn plant in Zhengzhou, China, a virtual high-tech city of 128,439 factory workers where the iPhone's final assembly[7] takes place (and where allegations about bad working conditions—some accurate, others fabricated—sparked a media sensation in 2012).[8] The finished iPhones are shipped to customers and retail locations in the U.S. and around the world to stores, cell phone service providers, and other outlets using virtually every transportation method known to man. Most of the U.S.-bound phones move by air freight through Hong Kong and Alaska, where UPS and Federal Express have major hubs. (The curvature of the earth makes Alaska a direct and ideal transshipment and fueling stop for air cargo moving from Asia to the U.S.)

This is the *partial* origin story of a collection of parts commonly known as the phone's home button, with about 12,000 miles required to get it to the place where the iPhone is assembled. All that is for one button, perhaps the least sexy part of a smartphone. And this triptych is just a partial accounting, because it does not include the movement of raw materials for individual components, nor their packaging, nor the movement of energy, water, and workers at the various factories, all of which could easily double or triple the mileage on that little button below the phone's touch screen.

Similarly epic journeys are attached to other parts of the iPhone: a barometric sensor and accelerometer from Germany; the Corning "Gorilla Glass" from Kentucky; the five different power amplifiers from California, Massachusetts, Colorado, North Carolina, and Pennsylvania manufacturers; the motion processors from Silicon Valley; the near field communication controller chip from the Netherlands; and many other components from Japan, Taiwan, Korea, and China.[9] The production of the Apple-branded A8 processor semiconductor chip is split between the world's largest contract chip fabricator, TSMC in Taiwan, and Samsung's immense new chip plant in Austin, Texas—a $9 billion investment by the South Korean technology company to make computer chips in the U.S. Samsung is offshoring *to* America.

Those parts, along with the Touch ID components, combine for that 160,000-mile commute embedded in the iPhone—two-thirds of the distance to the moon. And even that is still only part of the story. The movement of these components does not include the mining, processing, and shipping of the rare earth elements that are so vital to so much of our twenty-first-century technology, or the movement of the vast quantities of energy and water needed to obtain them.[10]

These materials, most with unpronounceable names that sound like minor Greek gods, are difficult to mine and pricey to extract from raw ore. Once refined, they can be worth many times their weight in gold. In recent years, China has dominated this rare earth market that the U.S. once led, though suppliers in California and Australia have been reclaiming market share of late. These "rare" materials—which are actually quite plentiful in the earth's crust, but rarely in sufficient concentrations to make mining practical—have almost magical magnetic, phosphorescent, and catalytic properties even in minute quantities. They are essential ingredients in everything from giant wind turbines and electric cars, to miniature electric motors, semiconductors, and rechargeable batteries of all stripes: phone-size, Tesla-size, and utility-scale–size. The iPhone contains a chorus of eight rare earth elements: neodymium, praseodymium, dysprosium, terbium, gadolinium, europium, lanthanum, and yttrium. These are not households names, but they are everywhere in the modern household, unseen yet invaluable. These elements can be found in a smartphone's color screen, various parts of the phone circuitry, the speakers, and the mechanism that causes a phone to vibrate when it receives a message or call.

Then there are the better-known precious metals inside each iPhone—a couple bucks' worth of gold, silver, platinum, and copper[11]—and the anodized aluminum enclosures. Together, the mining, refining, and transport of these materials—and all the chemical agents and systems needed to produce them—could easily double that 160,000-mile footprint on the iPhone (and any other high-tech product), as the precious metals, aluminum, and rare earths must be shipped from the sources to refineries and processors and then to the individual component makers around the world.

In the end, the iPhone has a transportation footprint at least as great as a 240,000-mile trip to the moon, and most or all of the way back. The wonder of this is compounded by the fact that this transportation intensity is a strategy to *increase* efficiency and lower cost.

On the face of it, that seems absurd. Yes, entrepreneurs and empires have traded goods across long distances for thousands of years: the ancient seafaring peoples of the Mediterranean, the merchants on the Silk Road, the classical Romans with their empire-wide network of paved highways larger than the U.S. Interstate Highway System. But back then, every mile added cost and risk. So only certain goods—rare fabrics, wine, art, jewelry, exotic foods, and bulk goods that simply could not be sourced locally—were worth global trading. Peppercorns were the original "black gold," not petroleum, and were once so rare and valuable that they stopped being merely prized goods and were used as currency, collateral, and even ransom in the ancient world.

The customer base for global goods in centuries past was almost as rare as the goods themselves. The vast majority of people until very recently consumed local and regional goods for most if not all of their needs. Sourcing everyday apparel, food, or common tools from distant locations—routine today—was out of the question and would have been outside the budget of all but a wealthy few. As much as such devices as the iPhone are the result of advanced design and engineering, they are also creatures of a supply chain that could only be imagined and afforded in this unique era of logistics and outsourcing in which traveling great distances for choice parts and processes is no longer a barrier. The real breakthrough that makes the iPhone possible—along with most of today's consumer goods, right down to the cheapest pair of boxers in your drawer or the salt-and-pepper shakers (and their contents) on your table—is a breakthrough of transportation.

It has not been this way very long. The dominant technology company of its era, the Radio Corporation of America, operated very differently. For decades after World War I, RCA defined the consumer electronic business with category-dominating products: radio, phonographs, and television (both as a manufacturer and a broadcaster). The company's labs also played a pivotal role in developing such other important technologies as radar, color television, the electron microscope, liquid crystal displays, and early computer systems. RCA grew to become one of the most recognizable brands and valuable companies in the world, investing its profits by scarfing up other companies and brands, including Random House publishing, Hertz car rental, and Banquet frozen foods (investments so far outside RCA's core competence that this "conglomerate strategy" eventually helped bring the company down). A promotional film, *The Reason Why*, details in classic 1959 newsreel deadpan how and why an RCA still at the top if its game made nearly every part of a television set in-house. RCA designed, prototyped, tested, and mass-produced all the major components for that era's most prized home technology: its vacuum tubes, the printed circuits, the cathode ray tube that formed the TV display, the tuners, the speakers, and even the finely finished, furniture-grade wooden cabinets that were the hallmarks of old-school TVs. In RCA's era the vertically integrated company—which controlled its supplies and components and thereby avoided paying another company's markup and the added cost of time, distance. and shipping—held a distinct advantage in the technology and consumer electronics sector.

A major competitor of RCA's, Motorola, pursued a similar strategy and was well-known for its early use of solid state components in place of vacuum tubes, culminating with its space-age-named "Quasar" line of TVs—an expertise that led to a new business: making computer microprocessors. Motorola, which designed the communications system used by astronaut Neil Armstrong to phone home from the moon, provided the processor cores for Apple's early computers as well. And Apple itself was much more vertically integrated in the past, sourcing components from other manufacturers (as the personal computer industry has done since its formative years) but also operating its own factories in the U.S.

Until the 1970s and 1980s, this strategy remained the norm for American industries. International trade of consumer goods was minuscule by today's standards,[12] often more trouble and expense than it was worth. The direct monetary cost of shipping, far higher in that era in real dollars, was just one barrier. Arrival times were unreliable, communication and tracking atrocious, and moving goods in and out of ports was notoriously slow and costly. Large gangs of longshoremen were needed to take cargo on and off every ship, with forklifts and nets but also by hand, in a painstaking style of cargo management known as "break-bulk" shipping, which differed little from the methods employed by ancient traders aeons ago. Cargo holds were loaded as a vacationer would pack a car trunk with suitcases, cramming in the various shapes and sizes of boxes and crates in vast jumbles. This was labor-intensive in both directions, slow, inefficient, and costly: a ship could spend ten days docked just to unload and reload. Combined with tariffs, breakage, and rampant cargo pilferage owing to so many loose items floating around holds and docks, international trade easily added 25 percent or more to the costs of consumer goods.

But then came the big breakthrough, a world-changing invention that would be both boon and disaster, making most of our modern and common products, not least the smartphone, possible. It was not some new ship design or propulsion system that launched the revolution, nor the advent of some exciting new high technology or manufacturing process. The breakthrough was as low-tech as could be: a steel box or, as American longshoremen call it, "the can." It's best known away from the docks as the shipping container.

In retrospect, the idea seems so simple, so obvious: Put everything in identical big metal boxes the size of semitrailers, stackable and uniform, each marked with a universal ID number. Make the containers the same everywhere in the world, design ships and docks specifically to accommodate them, and then sit back and watch the world change.

Simple or not, that innovation transformed—and exploded—global trade.[13] And it doomed companies such as RCA that could not adapt.

A big part of the magic was the fact that containers could be packed with anything and everything—televisions, furniture, coffee beans—*before* the ship that would carry them away even reached dock. The loaded containers could sit there and wait for the ship to come, not even requiring a warehouse to protect them from the elements: they are watertight. They could be sealed and locked to deter theft as well, the actual cargo never touched until the container arrived at the customer's doorstep to be unsealed. Tons of "containerized" goods could be piled on and off ships in one move, with a crane operator and a small ground crew instead of large gangs of longshoremen marching on and off ships, carrying a box at a time. Containers could then be placed right on semitrucks with empty chassis, with the already-packed container becoming the trailer. Or they could be stacked on flatbed rail stock: instant boxcars, fully loaded. Instead of spending more time docked while being loaded and unloaded than they spent sailing, the new breed of container ships were moving in and out of ports in a fraction of the time. Ships in motion make money. Ships sitting in port lose money. The value in time and money was apparent from the first.

In 1966, one of the first true container ships—the *Fairland*, owned by now-defunct America-based shipping giant Sea-Land—completed the first international container ship delivery, launching a successful weekly service between Port Elizabeth, New Jersey, and Rotterdam. The *Fairland*'s capacity of 236 containers was a tiny fraction of what modern container ships now carry, but it sparked the revolution nonetheless. The ships have grown spectacularly in size ever since.

The era of containerization coincided with a global movement to reduce tariffs to encourage free trade, and suddenly a $700 TV or a ton of iron ore cost only $10 to move halfway around the world. A $150 vacuum cleaner: $1. A $50 bottle of Scotch: 15 cents.[14] Containerization also came to air freight, and competition—in the air and on the ocean—drove down the cost of flying goods, particularly small ones, by an order of magnitude. This kind of pricing nullified the home-court advantage that companies such as RCA had enjoyed for so long by building the whole widget. The value of having a garment district in New York City or a tuna cannery in Los Angeles, a TV factory in Indiana or a vertically integrated car-making operation in Detroit, no longer seemed so compelling simply because it was close to the market for a product. The allure of low-wage workers eager for factory jobs in developing nations with little or no environmental regulation now seemed much more compelling as a source for car parts and computers and pretty much everything else. Although it was neither intended nor anticipated, the advent of containerization didn't just make shipping lines more efficient and profitable. Containers ushered in and made possible—perhaps inevitable—the modern era of offshoring.

At the same time, the complexity of new technology in the emerging digital age also undermined the RCA style of manufacturing. RCA had built an empire mastering the design and creation of mechanical, analog, and vacuum-tube technology. One of the company's first signature products, the phonograph, had been an entirely mechanical (and nonelectrical) device for a half century before electronic amplifiers and speakers were added to the hand-cranked devices. Significant parts of early TVs were mechanical as well. The first remote controls for TVs activated a motor that turned the mechanical tuner to another station with a series of impressive and loud *ka-chunks*.

The economics of making digital technology, however, were

different, requiring large investments in highly specialized fabrication equipment to forge processor chips, circuit boards, disk drives, and other complex components. It no longer made sense for a big electronics or technology company to invest in making the whole widget; the market favored factories focused on churning out large quantities of one major component to supply multiple and competing tech companies. The RCA do-it-all in-house style became a liability even as its technology became obsolete.

Soon the market leaders in American television and electronics manufacturing were exiting the business: first Motorola, and then RCA. Hand-polished wooden televisions had become just another plastic-encased commodity. With the magic of containerization ushering in a new era of globalization, Japanese companies took the U.S. electronics market by storm with such iconic products as the Sony Walkman (the iPod of its time) and the Trinitron TV, which set a new standard for image quality and reliability. In 1974, Motorola sold its Quasar brand and television business, once its most profitable endeavor, to Japan's Matsushita Electric Industrial Company, Ltd., maker of Panasonic electronics, and turned to the semiconductor business, which it later dumped to pursue the emerging cell phone business. The last of the old-school big American TV brands, Zenith, inventor of the modern remote control we all know and despise, sold off control of its brand and assets in 1995 to South Korea's LG Electronics.

The final piece needed to set the stage for globalization of consumer electronics and other goods was China's designation in 1979 of a "special economic zone" in a backwater town of 30,000 near the Hong Kong border called Shenzhen. It would be the first of six such zones, and foreign investment poured in, embracing the decision by China's communist government to make its country's immense and inexpensive rural workforce invaluable to the capitalist world. China's leaders had asked themselves a shrewd question: Why try to dominate the world—and bankrupt their country—with millions more men under arms when putting millions of men and women on factory lines and in engineering classes would tame world adversaries far more effectively and make the country rich to boot? Shenzhen today is a city crammed with skyscrapers, factories, and nearly 12 million residents, more than half of them migrant workers living in factory dormitories. Parts and products flowing out of China's special economic zones are ubiquitous now in America and the world, found on every aisle of every Walmart, every product category on Amazon.com, every device made by Apple, every car on the road, every restaurant, bar, power station, radio station, gas station, and train station; the list is book length itself.

When Apple CEO Steve Jobs hired Tim Cook in 1998 to run the company's worldwide operations—and eventually succeed him as CEO—it was not because of Cook's computer genius, but for his transportation acumen, his skills as a supply chain savant. Soon after his arrival at the company's Cupertino, California, headquarters, Cook proclaimed that Apple had to treat computers—then Apple's main product—like milk, a commodity that must be transported and sold quickly before it soured. This approach, now widespread, was made possible by the effects of the container revolution. Cook's goal was to have inventory cleared out in days instead of months, because idle inventory, like an idle cargo ship, is a costly drag on the bottom line. Such a strategy, Cook said, could never succeed with an Apple that owned its own factories. So he orchestrated their closure, along with most of the company's warehouses. Apple switched to a "just-in-time" manufacturing strategy that could only be achieved through outsourcing components and finishing products just days before they would be sold to Apple customers. The component suppliers, not Apple, would worry about inventory, but as they served many high-volume customers, their inventory cleared out far more quickly, which meant their parts manufacturing costs were lower than Apple's could ever be. In the era of containerization, the cost of sourcing across greater distances paled in comparison to the savings of manufacturing just in time. When Apple's most important lines of business shifted from a few million computers a year to tens of millions of iPods, then hundreds of millions of iPhones, this strategy paid off handsomely. Variations on this theme have transformed the entire consumer products industry, from toasters to sneakers to cereal.

So there it is: the three big disruptions that transformed the consumer goods industry and made one of the most popular gadgets—and most valuable companies—in history possible, bringing us to the morning of Friday the thirteenth and the sound of Big Ben bonging me awake.

There was the rise of digital technology. There was China's decision that, on the question of capitalism, it would rather switch than fight. And there was that most unsexy, nontechnological, big, and ugly metal can that spawned a transportation revolution. Of the three, it was the lowly shipping container—and its more spectacular progeny, the colossal container ship—that enabled the other two by turning massive amounts of transportation from a prohibitive cost into a transformative strategy. The container is both means and metaphor for this revolution, as the product it enabled—the signature product of the era, the smartphone—is the ultimate container itself, carrying inside it camera, calendar, navigator, reading library, music collection, transportation summoner—whatever we want it to be. The mundane iron shipping container spawned the supreme digital container, both of them innovations whose true impacts were unknown at the outset and are still evolving.

A defining quality of revolutionary change, however, is that its rewards last only until the next revolution comes along. RCA developed multiple revolutionary products to rule its analog roost for fifty years; now it's just an empty brand name licensed to makers of discount electronics. The current era of massive transportation footprints and distant outsourcing of nearly everything has also been in progress for fifty years, and once again forces are at work that could lead to profound change and less benefit for off-shoring. Call it . . . Cargogeddon.

The fleets of giant container ships that burn fuel not by the gallon but by the ton pose a growing environmental threat, with cargo vessels contributing about 3 percent of global carbon emissions now and on track to generate up to 14 percent of worldwide greenhouse gases by 2050.[15] But beyond their smokestacks, the mega-ships that now dominate cargo movement are threatening the transportation system itself, overloading ports and the networks of rail, road, and trucking that connect them to the rest of the world. The U.S. is running out of capacity at these choke points, with neither the money nor the will to increase it. The rise of online shopping is exacerbating the goods-movement overload, because shipping one product at a time to homes requires many more trips than delivering the same amount of goods en masse to stores. In yet another door-to-door paradox, the phenomenon of next-day and same-day delivery, while personally efficient and seductively convenient for consumers, is grossly inefficient for the transportation system at large.

And yet the impact of embedding ever larger amounts of transportation in products is often minimized in public discussion, even by businesses that have embraced the business case for sustainability. Certainly they are concerned about fuel efficiency in distribution and shipping—that's just good business—but the transportation footprint of a manufactured product is often a secondary concern at best. That's because the most common analysis of a consumer product's life-cycle—an estimate of its greenhouse gas footprint, which is a proxy for its energy costs—will usually find that the distribution of a product is a much smaller factor than its production. In its public disclosures on the footprint of its products, Apple states that transport accounts for only 4 percent of my iPhone 6 Plus's lifetime greenhouse gas emissions. Production of the device, meanwhile, accounts for 81 percent of its carbon footprint—twenty times the transportation footprint. Even my use of the phone—mostly by recharging it—overshadows shipping in Apple's life-cycle reckoning, producing 14 percent of its footprint.[16] For a glass of milk, shipping produces only 3 percent of the footprint.[17] For a bottle of California wine, it's about 13 percent.[18] Transportation accounts for only 1 percent of the carbon footprint of a jacket from eco-conscious Patagonia, Inc., even though it's made of fabric from China and sewn in Vietnam. Production of its petroleum-based synthetic polyester is said to be the main culprit, accounting for 71 percent of the garment's carbon emissions.[19]

These product-by-product analyses are accurate but often incomplete—and in the end, they can distort the reality of the gargantuan impact of the door-to-door system as a whole. Viewed as a sector, the transportation of people and product is second only to generating electricity in terms of energy use and greenhouse-gas emissions (consuming 26 percent of the country's total energy and fuel supplies,[20] while creating 31 percent of total greenhouse gases).[21] Transportation has a larger energy and carbon footprint than all the other economic sectors: residential, commercial, and agricultural, as well as the industrial/product manufacturing sector that figures so prominently in those life-cycle analyses.

Transportation leads all sectors in one unfortunate metric: when it comes to *wasting* energy, the movement from door to door tops every other human endeavor, squandering 79 percent of the energy and fuel it consumes.[22] Finding ways to reduce that waste presents one of the great economic and environmental opportunities of the age.

Wondering if this problem is about the movement of people in cars rather than products on trucks and trains? The simple answer: it's both. Proportionately, goods movement has the more intense carbon footprint in the transportation space, with transport by rail, truck, ship, and pipeline together generating about a third of the total transportation footprint. Freight trucks alone spew 22.8 percent of all transportation carbon emissions. Passenger cars account for 42.7 percent, while pickup trucks, vans, and SUVs contribute 17 percent. Given that there are fewer than 3 million big-rig freight-hauling trucks in America out of 265 million vehicles total,[23] the fossil-fuel-powered movement of goods has a disproportionately immense carbon, energy, and environmental footprint. Miles matter.

The attractions of offshoring are fading for other reasons, too. Conventional wisdom holds that cheap offshore labor is the main draw for American companies to shift operations abroad, but this grows less true each year. As a new middle class and consumer culture of its own emerges in China—one that wants to own iPhones as much as make them—wages have begun rising at a 20 percent annual clip (though admittedly from a low starting point, about $20 a day at suppliers such as Apple's Foxconn).[24] More important, labor costs have become a minor consideration regardless of location in the making of modern consumer electronics such as the iPhone—a tiny fraction of the overall production cost. Chinese labor accounts for only 1.8 percent of the price of an iPhone, researchers at the University of California at Irvine have found. And the total labor costs worldwide, including in the U.S., account for only 5.3 percent of the iPhone's price—almost inconsequential compared to the 21.9 percent Apple pays for materials to make the iPhone; and the largest piece of the pie, Apple profits, which were 58.5 percent when the study was done.

China, it turns out, makes little off the iPhone, while America—or, rather, an American company—makes a lot.

"Those who decry the decline of U.S. manufacturing too often point at the offshoring of assembly for electronics goods like the iPhone," the Irvine researchers wrote. "Our analysis makes clear that there is simply little value in electronics assembly."[25]

China keeps those jobs now because they have the very expensive facilities—and the engineering talent—to make products such as the iPhone in massive quantities. Silicon Valley is the world leader for designing these products and creating the software that makes them shine, but the Asian electronics powers have made themselves the nearly unassailable world leaders on actually building the stuff. As the previous CEO of Apple, Steve Jobs, told President Obama in 2010, America does not have that capacity. "Those jobs are not coming back," he said.[26]

But some of those jobs may disappear or be displaced in time as technology evolves—as the once-powerful RCA could attest, if the company still existed in any recognizable form. The advance of automation in factories, and the advent of 3-D printing as the next big thing in making consumer products, may usher in the next revolution and a new age of re-shoring. The future may belong once again to locally made or even made-at-home products: Buy the code for a pair of sunglasses or salad bowl online, the design is transmitted to a 3-D printer, and your purchase is created on the spot. Time will tell if the costs and capabilities of such technology will make it competitive. The only certainty is that the transportation piece of the puzzle is the one part that will never go away regardless of the tech. The direction of travel may change and miles may be shorn in the consumer goods space, but the transportation need will always be there, whether it is for finished products or the raw materials to make them. Manufacturing jobs come and go, but the logistics field just keeps growing—32 percent growth even during the Great Recession, while all other fields grew by a collective average of 1 percent.[27] Some say logistics *is* the new manufacturing.

Then there are those certain treasured products that will always have to be sourced from afar no matter what, and so always will pose a transportation challenge. One such good ranks among the most valuable and heavily traded commodities on the planet.

It's time to put down the phone and brew the morning coffee. But that's not the treasured commodity I touched next on this day in February, this Friday the thirteenth. That would be aluminum.

Chapter 2

THE GHOST IN THE CAN

My home is full of ghosts. I grab one parked on my nightstand before I go downstairs to make the coffee: an empty can of lime-flavored seltzer, a bedtime thirst quencher now destined for the recycling bin.

This pop-top metal can, a miracle of design that weighs half as much as a first-class letter, is emblazoned with the label "refreshe," all in lowercase letters. This is a ghost name for a ghost product. Every grocery, retail chain, and warehouse store has its ghost brands, their own private labels of products they sell but do not make. Washing machines, mouthwash, underwear, batteries, and every flavor of soda known to man—there are ghost versions of them all, their prices lower than the "name brands," their labels vague or silent on their true origins and journeys. It's all part of the vast puzzle of the modern consumer economy, in which we eat, guzzle, buy, and use things with no idea who makes them, where they come from, or what great dance of men and materials transports them to us.

But there are clues hidden in plain sight. My soda can is embedded with a coded road map that, properly read, lifts the sheet off the ghost of this ubiquitous, familiar object.

Most consumers don't notice or look for it, but each of the

billions of beverage cans made in America bears a small logo or company name unobtrusively tucked below the bar code or near the bottom rim. It may be an image of a crown or a company name apparently unrelated to the beverage inside: Ball and Rexam are the two most common ones, companies that began the merger process in 2015. These mark the maker of the can, not its contents. On the can's domed bottom—a shape not at all arbitrary but part of an intelligent design—is the familiar inked product "use by" date and code, the black machine print starkly obvious yet barely legible against the pale silver of the unadorned metal.

An even closer look yields information most never notice: two numbers a half inch tall embossed faintly into the can bottom rather than inked. These numbers, sometimes reversed as if in mirror image, are part of the quality control tracking process used by manufacturers of the cans. They identify the factory line and specific machine that turned a sheet of metal into a can. Before shipping, every can gets a fast but thorough inspection by computerized video in the time it takes a human to blink; the numbers let defects be pinpointed to an errant machine, metal die, alignment, or setting. It's no accident that every can is the same—these codes and inspections weed out the outliers.

Such clues also uncover the path a beverage takes from creation to expiration, one that explains much about the transportation embedded in all our products, lives, and choices, and how this particular container can be both very green and very dirty at the same time, ingenious yet also mad. The trail of numbers, logos, and names on my particular can of refreshe reveals a tangled path that begins in Australia but also includes most of the countries of Europe and the majority of states in America. Then it crosses oceans, ports, and freight yards to an aluminum mill in Tennessee, after which the material that will become my can of lime-flavored seltzer finally arrives at a rail spur in Northern California behind a beverage can manufacturing plant. From there it will travel by truck to a bottling plant and then a grocery distribution center for shipment to my local supermarket. That entire journey took as little as 60 days and spanned fewer than 10,000 miles for some of the material in that seltzer and its package. Other portions took as many as 60 years and several million miles before reaching my nightstand, an odd quirk of the door-to-door economy and the unique material that goes into that can.

Long or short, such a journey always starts with a reddish clod of rock and dirt called bauxite, inconveniently located in tropical regions of the world far from those who want it most. From that red dirt comes that most versatile of metals, aluminum, and a journey that touches every home and car and truck and plane—and soda can—in America, and the world.

Little cans are big business. America produces 94 billion of the beverage variety in a year, about a fifth of the world total. That's 2,981 aluminum cans pressed, cut, molded, and stretched into shape every second, which amounts to 293 cans a year of beer, soda, juice, and energy drinks for every man, woman, and child in the country.[1]

Aluminum is used in tens of thousands of products, from jetliner skins to automotive engine blocks, from glass and ceramics to thousands of miles of high-voltage power lines, from the liners of potato chip bags and juice pouches to the cladding inside nuclear reactors. From exquisite creations such as artificial sapphires to the mundane wrapping that keeps your leftover meat loaf fresh, aluminum has worked its way into daily life and the constant movement of ourselves and our stuff. This upstart metal went from zero to everywhere, from little known to essential in the space of little more than a century, while iron and steel needed a head start that predates Christianity by thousands of years to reach a comparably exalted position in our modern world. And no single product commands as much of the world supply of aluminum as the single-use, disposable beverage can.

One of every five pounds of aluminum processed in the world each year becomes a can for our beer, soda, and other refreshments. The lowly, ingenious can we barely give a single thought is also the single largest piece of a $90 billion global aluminum industry.[2] That means my seltzer can and all its little sisters are worth $18 billion a year before a drop of brew or bubbly water even touches it. Just on the basis of labor, energy, and material costs, my can of seltzer is worth far more than the beverage it holds.

On its face, this can may seem little more than a simple, convenient storage device, but storage is not the driver of this familiar object's design and worth, any more than a giant shipping container is designed for storage. The can's primary purpose is to enable the *transportation* of massive amounts of single-serve beverages more efficiently and cheaply than any other container type.[3] Its virtues include being lighter and stronger than reusable glass. When boxed, it wastes less space than tapered and necked single-use plastic bottles. And it is far more stackable than either of these rival materials when placed on pallets and loaded on ships and trucks, saving space, trips, fuel, and money. Think of the can as a miniature shipping container itself, designed to be placed easily in much larger containers and sent on its way. The fact that it stacks nicely in a home cupboard, refrigerator, or picnic cooler is merely a happy by-product of its shippability.

The aluminum beverage container, then, is as much a transportation game changer as those giant shipping containers that longshoremen also call cans, although the beverage variety is prized for more than its shippable design and weight. Beverage cans are prized most of all for the unique quality of the metal used to make them: its immortality. Alone among all manufactured substances, aluminum is infinitely recyclable. And, just as rare, it is highly profitable to do so.

This is how aluminum merchants can position their metal as the new transportation "killer app": a superlight, superstrong substance that never wears out at the molecular level but can be re-formed at will and sourced domestically over and over, thereby cutting production and transportation costs by a factor of ten. New supplies of the metal—what industry insiders call "primary" aluminum—are dirty and energy-intensive to obtain, after which the substance must be transported across vast distances to reach the majority of consumers. But once shipped, "secondary" aluminum can be continually reincarnated as a new product with many of the costs and distances stripped away. More than a century of mining has produced nearly a billion tons of the stuff, and an estimated three quarters of that remains in circulation and theoretically available for recycling from old cars, aircraft, appliances, obsolete TV antennas, and, of course, cans, which are recycled at far higher rates than any other single-use container.[4]

No wonder the beverage can has become the global poster child for recycling, the one "single-use" product that gets recycled more than it's landfilled.[5] The other big recyclables—paper and plastic—degrade during the recycling process, or lose value, or end up costing more than new material, so market forces for repurposing these waste products are mixed at best. Recycled aluminum, however, is a different story: not only is it chemically and physically indistinguishable from the new stuff, but it is beyond cost competitive. Aluminum recycling uses 92 percent less energy than mining and refining aluminum from bauxite,[6] and is often done near the end consumer rather than in far-off pit mines, lowering transportation costs and distance. Recycled aluminum is so valuable that its salvage earnings often underwrite municipal programs that recycle other more marginal plastic, glass, and paper waste.[7]

This explains why so much of the aluminum extracted from the earth since the 1880s is still in play, some of it recycled dozens or even hundreds of times. Some fraction of the aluminum in your car, your fridge, or your can of cola could have been mined more than a century ago. In previous incarnations it might have flown bombing raids in World War II, or made ice cubes in some 1960s refrigerator, or emerged from the thousands of tons of aluminum skyscraper sheathing salvaged from the destruction of the World Trade Center after 9/11. Or the can in your hand may have been someone else's can of soda as little as two months ago, because that's the current turnaround from fridge to recycling bin to factory to store shelf.

Kevin McKnight looks at such a can—or a new, lightweight wheel hub, or an experimental aluminum-based car battery that theoretically could boost an electric car to a 1,000-mile range—and sees the future. An aluminum-coated future, to be sure, which is to be expected, as McKnight works for the world's first and largest aluminum company, Alcoa. "We've reached an inflection point," the dapper Pittsburgh-based executive declared at an annual conference of environmentally conscious corporate leaders called Brainstorm Green. "The economics of aluminum are transforming whole industries, and transportation is our sweet spot."

McKnight's official title at Alcoa is chief sustainability officer, which puts him in charge of the company's efforts to become cleaner and greener—and, in the process, grow that transportation sweet spot into a mobility revolution. In practice his job might better be described as chief evangelist for an aluminum-rich future, and he is wildly enthusiastic about the possibilities as he travels the company's supply chain from Australia to Jamaica to Brazil and Canada, where he talks up the benefits of the industry his company invented at gatherings like Brainstorm.

Long the material of choice for aircraft and space vehicles because of its light weight and the fact that it does not rust like iron and steel, aluminum is now being touted as the next big thing for reinventing ground transportation. Aluminum is so light (atom by atom it weighs less than many gases) that swapping it with steel in cars and trucks could cut the average vehicle's weight in half, with corresponding decreases in fuel consumption and carbon emissions. In truth, such gains have been achievable since World War II. But McKnight's new pitch to carmakers—who have invested billions in steel bending and welding machines that they have been loath to replace—suddenly sounds much more enticing these days. The looming U.S. legal mandate that new cars more than double their average fuel efficiency by 2025 has seen to that.[8]

Turning cars into giant aluminum cans could go a long way to satisfying that goal, and McKnight and Alcoa, along with their competitors, have been pushing out new products to facilitate this "light-weighting" of vehicles to make them greener without necessarily abandoning the internal combustion engines that American carmakers (and American consumers) know and love best. Ford has now converted the body of its F–150 pickup truck—for three decades the most popular vehicle in America—from steel to an Alcoa-made military-grade aluminum alloy. This is just the visible thin skin of the truck body as opposed to the load-bearing structures beneath, but the partial changeover from steel still cut the truck's weight by 700 pounds. Given the 700,000 annual sales of the F–150, that's like taking 120,000 of the trucks off the road.

Vehicles are the single most recycled consumer product in the world,[9] so the increasing amounts of aluminum going into cars and trucks will eventually be recycled, providing the same 92 percent energy cost savings that beverage cans offer. But it takes nearly twelve years on average for passenger vehicles to enter the big recycling bin known as the scrapyard (and two or three times that for planes, trains, and cargo ships), with about 11.5 million vehicles scrapped annually in the U.S. Therein lies one of the great contradictions in the aluminum story and McKnight's sweet-spot pitch. Demand for aluminum in the transportation space has exploded—the record 504 million pounds of the metal delivered to automakers in 2014 is projected to rise to 2.68 billion pounds by 2018[10]—but recycling alone cannot yield the required supplies quickly enough. So ever more primary aluminum has to be mined and refined to meet the demand for more efficient cars.

This is how aluminum can be at once green and dirty, both a shining example of the "cradle-to-cradle" reuse economy and a coal-soaked, industrial-age relic of primitive extraction, spewing waste and toxins in its wake. This is where my can of seltzer's journey begins, straddling two worlds.

The world's largest bauxite mine is operated by Alcoa at Huntly in Western Australia, producing more than 20 million tons of the reddish brown ore each year. Towering excavators and loaders with wheels twice the height of their drivers, capable of hauling 190 tons of ore in a single load, crawl out of the pit in a constant stream. These dump trucks on steroids deliver the bauxite to the grinding roar of the rock crushers, which must be sprayed constantly with water to suppress the choking clouds of red dust spewing from their jaws. Boulders of ore torn fresh from the ground are pummeled into three-inch pebbles, each of which has 20 to 30 percent aluminum chemically locked within. The pebbles are loaded onto massive conveyor belts that snake down through forestland more than fourteen miles, delivering 5,400 tons every hour to Alcoa's sprawling Pinjarra refinery.

There the ore is put through a complex four-step chemical and cooking process, the main feature of which involves dissolving the material with copious amounts of caustic soda. This allows the ore to be separated into two streams, one rich in a precursor compound, aluminum hydroxide, the other, larger stream consisting of a noxious sludge called "red mud," which is shunted off to giant holding ponds. Red mud is the toxic albatross around the aluminum industry's collective neck, as there is no use or safe disposal method for the stuff.[11]

Heated to 1,100 degrees and treated with other chemicals in the next stage of manufacturing, the aluminum hydroxide sheds its hydrogen atoms and is converted to aluminum oxide crystals. When washed and dried, the material looks like granulated sugar but is hard enough to scratch glass. Commonly called alumina, these crystals are a little over half aluminum by weight and make for a convenient form for shipping in bulk. Together, alumina and bauxite are one of the most shipped substances on earth, one of what the shipping industry calls the "five major bulks." (The other four are iron ore, coal, grain, and phosphate rock, most of which is used to make fertilizer, a major U.S. export.)[12]

This method for refining bauxite was invented by the Austrian chemist Carl Josef Bayer (no relation to the aspirin inventor) in the late 1880s. It proved to be one of two critical discoveries needed to commercialize aluminum production. The Bayer process is still the only practical method for making alumina to this day.

From the refinery, alumina is shipped around the world, with about 10 percent of global supplies diverted for making such products as ceramic insulators, spark plugs, and other dense ceramics, as well as synthetic rubies for lasers and sapphire glass for watch faces (and, possibly, future smartphones). Alumina is also an ingredient in sunscreens and facial cosmetics. The stuff truly gets around.

Most alumina, however, goes to smelters—some near the refineries, some in the U.S. and across the globe—to be converted from sugary crystals into pure metallic aluminum. This is where the second critical process from the 1880s comes in, essentially unchanged for 130 years.

At the smelter, the alumina is dissolved in a molten mineral called cryolite, which possesses two amazing qualities: it puts the yellow in yellow fireworks, and it cuts the melting point of aluminum to less than half its normal 2,200 degrees, which means it also cuts energy consumption and cost. A synthetic version of cryolite is used these days, as natural supplies previously mined in Greenland have dwindled, and not from an excess of yellow fireworks. The molten mixture of alumina and fake cryolite is then placed in something like a giant battery cell, where it is jolted with fantastic amounts of electrical current via the process of electrolysis. The principles at work should be recognizable to every denizen of a high school chemistry lab who ever used a mild-mannered tabletop version of the process to break down a beaker of water into hydrogen and oxygen with a dry-cell battery and a pair of electrodes. In the same way, the cryolite-alumina mixture, liquefied at 980 degrees, is broken down by industrial-scale electrolysis, which unlocks alumina's chemical bonds joining oxygen to aluminum. The elemental aluminum that can't exist in nature on its own then sinks to the bottom of the molten mixture, a pure metal at last. Then the still liquid aluminum can be drained off, poured into long cylindrical molds, and cooled into ingots. The ingots used to make beverage cans run up to 24 feet long and weigh up to 46,000 pounds each. Each ingot can make 1.5 million 12-ounce cans.

Two different chemists working independently developed the process simultaneously: Charles Martin Hall in Ohio, and Paul Héroult in France. By the time the patents and lawsuits were settled and the smelting method became known as the Hall-Héroult Process,[13] Hall had founded a manufacturer he dubbed the Pittsburgh Reduction Company, invented the commercial aluminum industry, and become a billionaire. Later he renamed his outfit the Aluminum Company of America, which is now known as Alcoa.

These two processes from the 1880s are still used to make virtually all the aluminum in the world. Aluminum had only been observed in its pure metallic form a few decades before Bayer, Hall, and Héroult made their discoveries; until they came along, chemists were able to tease out only small amounts of this mysterious metal through laborious and expensive methods. Napoleon III, frustrated when he could not commission enough of the metal for a new generation of lightweight battle armor, settled for some very special aluminum dinner plates he reserved for his most esteemed guests; lesser visitors had to endure eating from far more ordinary (and less expensive) gold and silver. Before the French emperor's time, no one realized that aluminum was a metal at all, nor that it was the third most plentiful element on earth,[14] although various medical and textile uses for aluminum-rich compounds have been around since antiquity.[15]

In 2014, worldwide production of primary aluminum topped 53 million metric tons. Smelting that metal required nearly 690.170 gigawatts of electricity[16]—more than twice the power consumption of America's largest and most power-hungry state, California. Aluminum smelting uses more electricity than almost any other industrial process; engineers joke that the metal ought to be defined as "congealed electricity." Alcoa has located most of its smelting operations near sources of hydropower to lower the cost and environmental impact, but globally—particularly in China, with more than half the world's production—more aluminum is made with dirty coal-powered electricity than anything else. Domestic aluminum smelting in the U.S. alone consumes 5 percent of the electricity generated nationwide.

What this means is that aluminum's weight advantage over iron comes at a price: iron can be produced from iron oxide in a simple, relatively compact blast furnace; the complex Hall-Héroult process requires literally acres of electrolysis cells and city-scale power plants to produce equivalent amounts of aluminum. The bottom line: a car part made from steel costs 37 percent less than the same part made of aluminum,[17] although a life-cycle analysis by the Oak Ridge National Laboratory found that the overall energy and carbon footprint of a mostly aluminum car is less than a standard steel vehicle because of lower operating and fuel costs.[18] The calculation changes radically in aluminum's favor when recycled metal is used.

Once cooled, the aluminum ingots from which my seltzer can would be made were shipped out of Australia by cargo vessel to the Port of Long Beach, then taken by rail to Alcoa's Great Smoky Mountains fabrication plant in Tennessee. Ingots made from recycled cans are brought to the same plant. The metal used for cans is not pure aluminum but has small portions of magnesium and manganese (about 1 percent of each) mixed in for strength and stiffness, with the tops given an extra portion of magnesium and less manganese so it can withstand the stress of the pop-top. American beverage cans on average are 70 percent recycled metal, 30 percent primary aluminum.[19]

The Tennessee plant's main product is a five-mile-long coil of sheet aluminum used exclusively for beverage cans. Each 21-inch-thick metal bar is first heated and rolled into 3,000-foot-long coils an eighth of an inch thick, then cold-milled with massive rollers that bring the aluminum down to a thickness of a few thousandths of an inch. The Tennessee plant churns out enough of this thin aluminum sheet every minute to make 150,000 cans.[20]

The 7-foot-wide, 25,000-pound rolls of aluminum next travel by rail to an industrial park in Fairfield, California, to the Ball Metal Beverage Container Corporation plant. Headquartered in Colorado, Ball is one of those immense companies little known to consumers whose products are in so many homes; a $9 billion business with manufacturing plants worldwide, twenty-eight of them in the U.S. That makes Ball the largest beverage can maker in the world, churning out 50 billion cans a year for big soda and beer brands, as well as ghost-brand cans like those that contain refreshe.

The five-mile sheets of aluminum are uncoiled and fed into a cupping press, where rapid-fire blades strike home in a rhythmic, thudding rumble that fills the plant with a sound like a thousand marchers. The press cuts the metal sheets like a batch of silvery cookies, spewing out disks of aluminum several times wider than a finished can. These disks are then "drawn" through a die, meaning they are pushed through a metal doughnut by a cylindrical punch that forms each disk into a shallow cup about 3.5 inches in diameter. At this stage my can looks like a metal petri dish.

A conveyor belt moves this new army of cups to the next machine, the body maker. Each cup is pushed through a smaller die that squeezes it into the proper width of a twelve-ounce soda can, about 2.5 inches. The die forces the flexible aluminum to grow even thinner, the metal redistributed upward from thickness to tallness. At this stage, the can is not yet half its finished height, but it's getting there. This stage is called "redrawing."

The next stage pushes each cup through a series of increasingly narrow dies that stretch the cup gradually to its proper height while making the walls thinner—like stretching a rubber band, except the metal doesn't snap back. This is called "ironing."

At the end of the ironing, a dome-shaped die is pressed into the bottom of each can. Whether used in buildings or cans, an arch or dome is stronger and can withstand more pressure than a flat surface. Doming allows the can bottom to be thinner, saving material, weight, and money. Each of the dozens of doming tools on the assembly lines has a unique two-digit number, and those numbers are embossed on the can bottom when the dome shape is pressed into place.

All this happens fast: the redrawing, ironing, and doming of a can takes about one-seventh of a second.

Next the can's open top end is trimmed for a clean edge and uniform height, followed by multiple high-temperature washes, drying, and the painting of labels. The can is baked to harden this "decoration," as the labels are called, followed by a spray of waterproof varnish inside the can to keep beverages from tasting of aluminum and to protect the can from reacting with acids in a beverage.

The narrow neck of the can is then formed by passing it through a series of eleven "necking sleeves," then the top is folded over into a flange; the can top can be attached later. The neck (early aluminum cans had little or no necking) is not for aesthetics: it reduces the amount of aluminum needed for each can slightly, reducing its weight as well as cost. While inconsequential for one can, the effect across 100 billion cans is massive: the current amount of necking saves about 100,000 tons of aluminum a year. That's enough to make a solid cube of aluminum 105 feet tall—higher than a ten-story building.

Computerized video cameras scan each can for leaks and imperfections, then the finished cans, their mouths gaping open, are packed on pallets and wrapped in plastic film to hold them in place for shipment to the beverage plant.

A separate machine stamps out the can lids with integrated

pop-tops at a rate of 6,000 a minute. The pull key that opens the can looks like it's riveted in place, but there is no separate rivet; that would break the seal of the can and allow leaks. Instead, the shape of a rivet is drawn out of the aluminum lid material itself, providing a seamless flange that holds the separately made pull key in place with a fold of metal.

The cans and lids are then shipped to the Safeway supermarket chain's bottling plant thirty-two miles away in the San Francisco Bay Area city of Richmond, California. On the refreshe line, ordinary water is mixed with natural lime concentrate and injected with carbon dioxide to give the seltzer its fizz. As soon as the cans are filled, the bottling machinery attaches the lids by folding the metal twice into a double airtight seam without welding or solder—just a little liquid gel inside the folds that hardens into a gasket to prevent even microscopic leaks.

The carbonation inside the can causes twice normal atmospheric pressures. This is why aluminum cans can be so thin: the pressure inside is always pushing out against the walls of the can, making the structure stronger and very difficult to squeeze or damage. This is why crushing a sealed can of soda in your hand is impossible, and why crushing an open can is child's play. This is also why even non-carbonated teas, coffees, juices, and other canned beverages hiss when opened. They are pressurized as well as the top is slammed into place, although inert nitrogen, not carbon dioxide, is used for these drinks. Nitrogen does not make drinks fizzy.

My seltzer left the bottling plant in a twelve-pack carton, entered the Safeway distribution system, and found its way by semitruck to my local market, a subsidiary of the Safeway company called Pavilions. The pack would go on sale for the ridiculously low price of $2.49, the true magic of the ghost brand being its low cost. The next time I would purchase some, the can would show me a different road map, having been fabricated by Ball's arch can-making rival in the fiercely competitive aluminum business, Rexam,[21] at a plant in in the Los Angeles area, then bottled at Safeway's Norwalk plant, a town just ten minutes from my home.

When enough cans have accumulated in my garage—okay, when too many cans have accumulated in my garage so we can't procrastinate anymore—we bring them to the recycling outpost in back of that same supermarket where I purchased them. Because of California's robust container deposit law, we receive a dime refund for every can we turn in, one reason why the state is the national recycling leader. Only ten states impose container deposits on beverages, however, and this explains why, nationwide, America's recycling rate compares unfavorably with Europe's and Japan's. It's also why, despite the value of scrap aluminum, 43 percent of aluminum cans used by consumers still end up thrown away instead of recycled.[22]

As a consequence, the only way can makers can achieve the 70 percent recycled content in U.S. soda cans is by importing old cans from elsewhere in the world, mostly Europe. And so the metal in my can of lime seltzer—and every other canned beverage in America—is far better traveled than most of the consumers who buy it, as the industry is forced to outsource the metal from old cans from around the globe to satisfy our thirst. The cost of hauling scrap aluminum cans around the planet might knock some of the shine off the industry's green credentials, but it still pencils out: even old cans transported from abroad are cheaper and have a lower energy and carbon footprint than pulling that same metal out of the mines.

The technology, ingenuity, and massive amounts of transportation embedded in my simple canned carbonated beverage is in many ways a perfect case study, a microcosm for our entire way of life, commerce, and movement. No one company nor any one country could make the whole widget, from Australian ore to lime flavor. It's real lime, not artificial, the label on my can of refreshe says, so someone had to grow the limes; fertilize, water, and pick them; then squeeze, package, and ship the juice. Someone else had to make the paint and varnish for decorating the cans, and package those products and transport them. Someone else had to make that carton for the cans and the wooden pallets they are shipped on, and the plastic shrink wrap that keeps the featherweight containers stacked instead of flying around inside the truck. It is a true global product, this ghost-brand soda, one that could only be made in this way in today's door-to-door world.

For all its wonder and power, though, a trap lies hidden within this door-to-door prowess. This seamless behind-the-scenes delivery machine can magic a red rock in Australia into a seltzer can in my refrigerator with utter consistency and reliability, and if that's not remarkable enough, sixty days after I drain it and toss it away, it can be back in my refrigerator again, good as new. This is an extraordinary product resurrection. Yet habit, time, and ubiquity have drained this achievement of its true wonder and rendered it not just ordinary but beneath our notice. It's simply a thing we buy and use and expect, which is the unintended but inevitable accomplishment of our modern, have-it-now logistics age: turning the remarkable into the mundane. And that's the trap: Who questions what's beneath notice? Who asks why—if—we need such products, or even if they make sense?

Instead of questioning the very nature of the can—or the ship or the car or any other staple of the door-to-door world that has become part of daily American culture—the focus is almost always on *refining* the magic. Make cargo ships twice as big in the space of ten years so they can carry even more stuff door to door—but give no thought to the impact on roads, traffic, and infrastructure when all this extra cargo slams into land. Or make cars lighter with aluminum so they burn less gas and emit less carbon. But don't question the transportation fundamentals these lighter cars will perpetuate—a country where 57 percent of households own two or more cars,[23] all of them spending an average of twenty-two hours a day parked and disused.

And then there's the can. In 1972, the most brilliant manufacturers on the planet managed to squeeze twenty-two cans out of a pound of aluminum. Now the industry standard is thirty-four. Researchers are currently working on eking out a few more. Perhaps they'll reach forty cans a pound someday. Think of the aluminum, the energy, the resources, that could be saved, the industry spokesmen say, and they are absolutely correct. But at the end of the day, it's still a can that we use once and throw away, *sometimes* into a recycling bin. It can't even be resealed once opened. How many quarter-full cans of flat soda or beer are spilled down the drain, wasted because that is the intention behind the design? No one knows for sure, but even with a conservative estimate of 5 percent of canned beverages wasted, that would amount to 9 million gallons a year—all that weight and energy moved across cities and continents just to be thrown out. Is this really the most efficient and sensible solution that our geniuses of movement and design can come up with? You could ask the same sort of question of cargo ships or cars, of course, but just imagining American households without car ownership or American ports without massive imports is a tough sell in a landscape and economy designed around them.

But the can shows the way. History, rather than imagination, reveals an alternative reality. Soda water, seltzer water, club soda—whatever the chosen name—carbonated water has been around since the eighteenth century, long before the aluminum can came along, or industrial bottling, for that matter. A soda siphon, seltzer bottle, or other reusable device for carbonating water—early versions used bicarbonate of soda as the carbon dioxide source, later versions featured injections from small canisters of the gas under pressure—was a common household implement from the late 1800s through the 1930s. People would drink it straight as a health tonic or mix it with other flavors, usually alcoholic ones. And when Coca-Cola became the first commercially popular soda pop in the U.S., it was purchased as a concentrated syrup. Consumers could make their own Coke at home or, more commonly, go to a drugstore soda fountain to be served a hand-mixed treat, ice cream optional. The notable feature of either alternative was a lack of single-use containers. Consumers could make their own plain soda and add the flavor sold by Coke, its competitors, or additives of their own devising. One purchase of syrup in one container made many servings, a system that was efficient, low-waste, and needed minimal logistics to close the deal. The heaviest and least expensive ingredient in a single serving of soda—the water—was provided by the consumer, so it didn't have to be bottled, canned, or shipped. If there must be mass consumption of a nutritionally deficient, obesity-causing sweet, fizzy beverage, there is no more efficient or low-cost method for distributing it than this original model.

But it is not the most *profitable* model. Taking its cue from the beer industry, which crafts a product not so easily made at home and therefore in need of a bottle or can, the soda industry marketed a new innovation: single-serve, ready-to-drink glass bottles, which led eventually to plastic bottles and aluminum cans. Shipping all the extra water and weight was a challenge but well worth it from the industry's point of view: there was so much money to be made selling many single-serve containers to a customer instead of one container of concentrated syrup. You bought a soda siphon only once and a supply of syrup only infrequently. But single-serve containers have to be bought over and over, and from the industry's point of view 95 percent of the drink being sold is simply water (99 percent for diet sodas). The most costly parts end up being the container and the transportation.

From an efficiency, shipping, and waste point of view, this shift made no sense. Consumers were paying more to get less. But it was marketed as an innovation, as progress, as convenience. And that fussy old siphon belonged to an older generation. Drugstore soda fountains, which had gained in popularity when Prohibition shut down the bars, became passé. The market—and the marketing—spoke. Now, a century later, there is a slowly growing niche for home soda makers offering a back-to-the-future appeal and a more sustainable model, but America for the most part wants its soft drinks in disposable, single-serve containers. The can business is booming. But the history of the can reveals this to have been a choice, not an inevitability—one that has been profitable for a few enterprises, but costly for consumers as well as for the planet.

When considering our most popular, enduring, and costly container, the car, and how inevitable its current form has seemed since World War II, it may be useful to think of it, too, as a great big can: a choice, not an inevitability, for the future of the door-to-door world.

Chapter 3

MORNING BREW

"The Industrial Revolution absolutely ruined coffee," Jay Isais is saying over the basso roar and BB-pellet clang of his coffee roasting and packing plant. "Entire generations grew up drinking an absolutely terrible product you could only tolerate by masking the taste with lots of milk and sugar."

Isais is showing off his caffeinated domain as he speaks, a business launched back in 1963 on the then-novel notion that java could and should be better than the traditional big cans of brown grounds on the supermarket shelves. As he watches, the latest batch of Costa Rican Tarrazu tumbles out of one of the giant Probat roasting machines with a sound like pounding rain, 600 pounds of beans cooked to 400 degrees for 13 minutes, give or take a few very critical seconds. The roast time varies from one coffee variety to the next, and it doesn't take too many seconds over the limit to ruin a batch.

Not today, though: a lovely hot, sweetish, overpowering aroma of coffee wafts from the stainless steel cooling platform where the beans just landed, its rotating arms sweeping them around and around as if panning for gold, lowering the coffee temperature enough to be moved to either the grinder or the automated packing station for whole bean customers. At the same time, a technician grinds and scans a sample of this batch with an ultraviolet sensor. He must make sure the beans have achieved the right color of a proper medium roast, not too light and not too dark, cast-iron low tech meeting computer-age refinement in the quest for a truly good cup of java.

Jay Isais is nodding and smiling as the readout comes within a percentage point of the target. He is an unabashed coffee nerd who also happens to run sourcing and manufacturing for the biggest coffee house chain in the U.S. not named Starbucks. He's the Coffee Bean & Tea Leaf's senior director of coffee, roasting, and manufacturing—-or, in lay terms, the company coffee guy. He literally lives, breathes, and slurps coffee for a living: the company has nearly a thousand stores in thirty countries, and every one of the 8 million pounds a year the company buys is personally chosen by Isais. He oversees the whole convoluted supply chain from field to ship to roasting to store, all centered in the company's single roasting plant in little Camarillo, California.

"This would be a tough job if you didn't love coffee," he allows, not that there's much doubt about where his heart lies when it comes to America's most frequent daily sip.[1] His office is a monument to coffee. The pictures on display are primarily group shots of Isais with coffee farmers he has visited and worked with in developing countries. His credentials include being a founding member of the Roasters Guild, a volunteer instructor for the Specialty Coffee Association of America, a certified judge for various coffee competitions and organizations, a certified supply chain professional specializing in coffee and tea, and a licensed "Q Grader"—the coffee industry's anointment for experts in the art and science of tasting and judging the qualities of coffee. There are about a thousand Q Graders worldwide. Isais talks about coffee as a spiritual experience as well as a business, and it's clear he takes bad coffee as a personal affront.

This is what bugs him about the legacy of the Industrial Revolution. It worked wonders for the advancement of cars, plumbing, and electricity, he says. But it wreaked havoc on our cups of coffee.

What most consumers don't realize, Isais says, is that when they buy coffee in a big can at the supermarket, it's already stale before the first cup is brewed—even before the can is opened with its impressive hiss of a vacuum seal released. This is simple chemistry at work: along with its delicious aromas, coffee gives off copious amounts of carbon dioxide for a day or two after leaving the roaster. Stick the java right in a can, and that can will begin to bulge or even rupture from the pent-up gas pressure. Wait until the outgassing slows before sealing the can, and the problem goes away—but so does freshness. This had been the problem with American coffee since early in the twentieth century, when mass production and canning techniques were first applied to what had previously been a commodity sold fresh or even raw to the public.

This seems like a packaging problem, but at root, Isais says, it's a problem of transportation and of a supply chain that, in its own way, is as complex as a smartphone's. How do you handle a highly prized commodity that grows only in certain tropical locations at very specific altitudes, on millions of mostly small family farms, none of which are near the bulk of the customers who want to buy and drink the stuff? What do you do with a product that is highly perishable when it's picked, that can be partially processed to a raw, green stable state that will keep for many months, but that becomes highly perishable again once it's roasted and ready to brew and drink?

Unless individual consumers can take the time and trouble to roast the green beans themselves *and* use them right away, or go to a coffee shop and drink brew made from recently roasted beans whenever they want a cup, everything about the taste of coffee is going to be a compromise between convenience, freshness, distance, and time. In other words, it's about transportation.

"Many people never know what coffee is supposed to taste like," Isais says. He has a lean, expressive face framed by a closely trimmed beard and mustache that can't conceal his pity for the 85 percent of Americans who drink at least an occasional cup, and the 63 percent who drink it daily.[2] A majority of these regular coffee drinkers are making stale brew and think it's supposed to be that way. He vividly recalls the first time he tasted really good coffee. He was a college junior starting what was supposed to be a temporary job at a family friend's coffee roasting business in the Northern California coastal town of Monterey. "It was a revelation. Coffee wasn't just that muddy brown stuff that came in cans that you had to dump a ton of milk and sugar in, but something amazing."

Before the mass production techniques Henry Ford brought to the automobile were applied to coffee, the product was most often sold in its raw green bean state in the U.S.—the beans having been cleansed of the fruit skin, pulp, and an inner husk called the parchment, but not roasted. Coffee can stay good for up to a year in this state if kept dry and indoors. Consumers would take it home, roast it in a pan or oven, and grind it with a hand-cranked coffee grinder. The drink became somewhat popular in America during the American Revolution. Patriots wanted to supplant their previous favorite, tea, after the Boston Tea Party. Serving coffee represented a statement against British custom and rule. But coffee really took off as an American staple nearly a century later, during the Civil War. It was one of the few luxuries—as well as a welcome stimulant—offered troops on both sides, although only the Union Army had reliable supplies after the first year at war. Hundreds of thousands of men came home from the war hooked on java. Green coffee beans were part of the daily rations given to Union soldiers, who had little roasting kits in their packs or just used cast-iron skillets on the campfires. Some of the government-issue carbines had little grinders cleverly built into the rifle butts, but others just used their regular, solid rifle butts to hammer the beans until they broke up enough to brew.

Isais finds this bit of history fascinating and illuminating, because America fell in love with good coffee in those days. "The irony is those Union soldiers were drinking better coffee out there in the field than any fine diner was served in the best restaurants in America in the 1950s."

That difference not only hooked Isais into a career, but also powered the modern coffee industry to new heights. Then it split the industry in two. Today there is the commercial coffee, canned and least costly, that nearly everyone in the U.S. served and drank up until the seventies, even at the finest restaurants. And there is the specialty coffee business, jump-started by the founder of Isais's company in 1963, and catapulted into ubiquity by one dominant player, Starbucks, with its mega-fleet of 32,000 stores. Now that specialty branch is forking again, pushed into the gourmet beverage stratosphere by so-called third-wave artisanal coffeehouses whose product is sourced, swished, savored, and scored like fine wines.

Divided or not, coffee is big business—with $28.6 billion in worldwide exports alone.[3] Those exports generate at least five times that amount in resales by the pound and by the cup, in a world where 1.4 billion cups of coffee are consumed daily.[4]

However, contrary to oft-repeated claims in print and online, coffee is not the second most valuable traded commodity on the planet after crude oil. This bit of lore had been accepted and repeated for years, perhaps because it resonates so powerfully, this idea that there are two great black elixirs in the world, oil and coffee, that keep us moving, albeit in different ways. But even the most rudimentary research reveals that coffee is not even the second most valuable *agricultural* commodity traded globally, much less in comparison to such mega-products as oil, gas, coal, alumina, and all the metals and gems that are traded in the world. Coffee is, however, among the top ten agricultural imports in the world—number eight, to be exact—behind such products as soy, wheat, maize, rubber, and wine but ahead of beef, cigarettes, cheese, and rice.[5]

None of the countries that produce coffee are among the top per capita consumers of the beverage, except Brazil, which is in tenth place worldwide. And although the U.S. consumes the greatest volume of coffee of any country in the world—second only to the collective imports of the entire European Union—Americans are way down in the pack when it comes to how much coffee *per person* we imbibe. The U.S. is in twenty-second place on that list, tied with Poland and Hungary. The Scandinavians are the coffee-drinking world champs, with Finland taking the caffeine crown by consuming more than twenty-one pounds of coffee a year—about two and two-thirds cups a day—for every man, woman, and child in the country. That's more than three times the per capita consumption in the U.S.

If all that coffee could grow anywhere on earth, Jay Isais's life would be a great deal simpler, as would the challenge of keeping coffee fresh and delicious for people from Finland to Florida. But it cannot. Coffee is a tropical shrub adapted to a specific climate that rules out growing coffee where its main consumers live: in Europe and North America. Every U.S. state consumes the stuff, but the only state that actually grows it is Hawaii, and its volume is barely a blip on the world's coffee radar, although its Kona coffee beans are both famous and famously expensive. And no traded coffee at all is grown in Europe.

So coffee requires massive amounts of transportation to reach the world's java guzzlers—not just from the producing countries to the consuming countries, but through a web of intermediate processor, broker, and marketing nations. Germany, without a single commercial coffee tree on its soil, is in the top five coffee exporting nations in the world, acting as broker for other nations and also as the world leader in making decaffeinated coffee through a process that removes the stimulant but leaves the coffee in its raw green bean state. The extracted caffeine is then sold, primarily to the pharmaceutical industry, and the beans are shipped out again. Five percent of all U.S. coffee imports come directly from Germany—more than from the coffee nations of Costa Rica, Peru, and Nicaragua.[6]

All told, more than 142 million bags of coffee, each weighing 132 pounds, are moving each year through a complex dance of ports and countries.[7] Each coffee plant produces about a pound of raw green coffee beans a year, so it takes a great many plants and a great many pickers to keep the world's cups full.

The coffee I'm brewing on Friday the thirteenth is a medium roast from the Ethiopian village of Yirgacheffe, in the equatorial highlands of the Sidamo region, where local growers believe the first coffee plant originated and where ancient methods of processing are still used. Whether or not this place really is ground zero for all coffee on the planet has never, and may never, be proven. But there's no question it produces some highly prized beans.

My coffee's physical journey starts when the beans leave Yirgacheffe on a rugged 250-mile ride to the capital city of Addis Ababa for final processing and packing. Ethiopia is landlocked, so the next leg in the coffee itinerary is another 536-mile drive across the border to the bustling commercial port in the Republic of Djibouti. From there the beans travel by ship north through the length of the Red Sea and the Suez Canal to the Mediterranean Sea. They cross the Mediterranean's entire length, then pass through the Straits of Gibraltar to the Atlantic Ocean. Typically there would be numerous port calls along the way. Then the ship heads west and south again to the Panama Canal and passage to the Pacific Ocean, where it turns north and west to California and Jay Isais's waiting hands.

Assuming the coffee travels the most direct route, that's about 11,000 nautical miles. (By contrast, a coffee shipment from number one producer Brazil to New York would be just under 5,000 miles at sea.) But it's not usual for coffee to stay at sea two months before reaching its final destination, as ships often take roundabout routes to ports of call for pickups and drop-offs in the opposite direction, so the final mileage could end up being much more.

Coffee blends—different varieties shipped separately and blended in the consuming nation after roasting—can double or triple mileage as well, depending on the sourcing. Most commercial coffees sold in the U.S. are blends that include beans from both Latin America and Asia, and many popular specialty coffee sellers blend beans as well, including some of Starbucks' most popular coffees. The upshot: the average American coffee-drinking household (with two moderate coffee consumers using blended bulk ground or whole bean coffee) never has fewer than 572,000 miles of travel pass through its coffeemaker every year.[8] My household, which consumes java at a rate more in line with Finland's, pencils out at about 1.7 million miles—and that's just counting the transportation from farm to first arrival on U.S. soil. There's also the added travel from port to roaster to distributor to grocery or coffeehouse. There are still more miles if the household buys decaf or uses a single-serve coffee machine, for which many of the coffee capsules—Nespresso's, for one—are sourced from Europe.

This massive coffee industry started very small in northern Africa, probably somewhere in Ethiopia's southwestern highlands, though not necessarily Yirgacheffe. DNA analysis suggests this general area is where *Coffea arabica* first appeared, sometime between 500,000 and one million years ago.[9] Coffee existed long before there were any modern humans around to appreciate this flowering evergreen with shiny leaves and fruit resembling (in appearance, not taste) a brilliant red cherry. Inside the cherry lies the coffee bean, which is not a bean at all but two semicircular seeds.

Coffea evolved from a large family of plants known as Rubiaceae, whose "tribes" also include gardenias, hydrangeas, and plants that produce key ingredients for important medicines, such as quinine, the first effective treatment for malaria, and coumarin, used in the anticoagulant drug warfarin. The only real "food" in the group—although it contains almost no calories when brewed—is *Coffea*, of which there are numerous types, the most prized and commonly used being *arabica*.

Legend surrounds the discovery of the delicious properties of the *arabica* seeds, which could not reveal their secret until properly prepared and roasted. The most beloved origin story is the Arabian tale of a clever young goatherd named Kaldi, who one day observed how surprisingly peppy his goats became after munching on the fruit of the coffee plant. The excited boy went to the nearest monastery, showed the fruit to one of the monks, and told him of its amazing effects. The dubious holy man threw the stuff into the fire, only to be captivated by the aroma emanating from the burning fruit. The holy man and his companions then decided to place the roasted beans into boiling water as if making tea, and so coffee was born—at least in legend. The tale of Kaldi and the holy man has a multitude of versions and all are almost certainly mythical, given that there was no recorded account of the Kaldi story until about eight hundred years after the events were said to have taken place. But the charming origin tale has endured.

The first historically reliable account of anyone drinking roasted and brewed coffee as we know it today dates back to the fifteenth century in the Yemen region of the Arabian Peninsula, where it was used in religious devotions in Sufi monasteries.[10] But coffee was just too good for the monks to keep to themselves, and it soon spread throughout Arabia, then Persia, Turkey, and every other part of the Middle East, where coffeehouses started springing up throughout the Islamic world like Middle Ages Starbucks. Within a century, the enterprising traders of Venice had brought coffee to Europe, where it survived an attempted ban as a heathen evil after the pope tried some and gave it an official papal thumbs-up in 1600. With a new coffee trade thriving between growers in Africa and importers in Europe, Dutch traders found the plant could be grown in Java and Ceylon (now Indonesia and Sri Lanka) as well. Not to be outdone, the French brought it to their New World colonies, and coffee plantations soon sprang up in the Caribbean. Next the *arabica* plants gained a foothold in South America, where they thrived, and in time Brazil and Colombia would become leading coffee-growing countries. A long history of colonialism, slavery, and exploitation ensued, and the exploitation persisted for centuries. Only relatively recently, through pressure for industry reforms and the fair trade movement, have conditions and compensation begun to improve for the estimated 25 million family farmers and producers in 50 countries who subsist on coffee.

The *arabica* coffee plant flourishes in tropical climates at altitudes ranging from 3,500 to 5,500 feet above sea level, with the best results in hilly areas with partial shade. The majority of beans are handpicked, with each plant requiring multiple pickings because the ripening comes in stages over the course of six to eight weeks. Poorer grades are just swept up at once, picked by machine where the terrain allows it.

The freshly picked coffee cherries have to be processed immediately, as the outer fruit rots very quickly, degrading the flavor and quality of the beans within. This is the part of the process that makes or breaks good coffee, and Isais travels each year to visit his source farms to see how it's done and if it's done right. For *arabica* beans, the most common procedure—called the wet process—requires immersing the beans in open water tanks resembling swimming pools, where natural fermentation causes the fruit to slough off the beans. Then the beans are dried, a task accomplished on small farms by spreading the wet beans on outdoor patios or platforms, turning them occasionally by hand. Large estates employ huge tumbling commercial dryers for the task. Once dry, a final mechanical hulling removes the thin parchment that encases the seeds, somewhat like the papery skin inside a peanut. Some farmers—those in Yirgacheffe among them—still use the ancient "dry method," by which the fruit is allowed to dry like a raisin in the sun, then is removed from the beans through a milling process. It's harder to get consistent results this way, but when done with care, this sort of coffee is valued for its very complex and intense flavors.

All of this work typically takes place in the coffee's source country, usually near the farm, and in some areas even the smallest family farms do the processing themselves or in small cooperatives. Other growers rely on local businesses to buy the freshly picked cherries and process batches of local beans from many growers. But this is the one stage where coffee doesn't wander very far from its tree. Adding too many travel miles at this stage would take too much time and reduce the grade of the coffee and therefore its price. So in the coffee world at least, the ancient rules hold for the freshly picked fruit: time and distance is the enemy, and no technology or shipping container or shiny outsourced factory can stop the spoilage and beat the clock.

After processing, the beans are separated into as many as fourteen grades based on size, consistency, appearance, and quality, with the top grades going to the specialty coffee buyers such as Isais, and the rest to commercial buyers or instant coffee factories at a much lower price. The bottom grades are too poor to be exported and are consumed locally, if at all.

There are dozens of people making decisions about how to handle each of the stages of gathering and processing, and an entire coffee lot can be ruined by any one of them. "That's the truth of coffee," Isais says. "It can never be made any better, but there are a thousand ways to make it worse along the way."

When the coffee emerges from the final milling as raw green beans, the rules governed by coffee's fragility change: the green beans are sturdy and can last months, even a full year, with little or no care required. The green beans pass to agents and brokers who aggregate coffee from small lots in a region and sell to exporters, who deal with the coffee buyers and their brokers and agents worldwide. It's a convoluted process, and it's not unusual for coffee to pass through many ownership changes before reaching a consumer. Sometimes coffee can be purchased directly without all the middlemen, and this "estate coffee" is sold like a premium wine. The Coffee Bean & Tea Leaf has been buying directly from the same coffee agent in Yirgacheffe for fifty years, Isais says, allowing it to obtain beans from the same estate of small landholders in the village near the birthplace of coffee.

At the other end of the quality spectrum is a different variety of coffee, made from the plant *Coffea canephora* (or *Coffea robusta*), commonly called *robusta* coffee. It's grown primarily in Brazil and a relative newcomer to the coffee game, Vietnam, which has no significant *arabica* production. As the name implies, *robusta* is a heartier plant than *arabica*, less fussy about altitude and climate, perfectly happy to grow in flatlands where mechanical picking is easy and cheap, and it even has 50 percent more caffeine. *Robusta* would be the perfect coffee for the modern world but for one thing: the taste. Most people find it bitter, biting, and far less appealing than *arabica* coffee.

But like bad wine, *robusta* has its attractions: low cost. Its bargain prices make it a preferred choice for instant coffees, compensating for the added cost of the instant coffee process, which does not ruin coffee flavors in itself. The big commercial brands often mix *robusta* along with *arabica* in their canned coffee blends (the precise blend ratios are kept secret). Again, this is a cost savings that keeps the price of canned coffee down. According to Isais, the flavor suffers from another cost savings as well: the practice of roasting commercial coffee at a lower temperature in order to avoid losing as little weight as possible as the beans cook and dry out. A pound of under-roasted coffee takes fewer beans than a pound of fully roasted. All this makes for far more affordable coffee, but it's also part of the reason why specialty coffees seem to taste so much better than the old-school canned varieties.

The coffee supply in America is a broad mix. If all coffee, *robusta* and *arabica* alike, from the top ten coffee-growing nations that supply the U.S. market were made into one big blend—call it the American House Blend—this would be the recipe:

Brazil:	29 percent
Columbia:	18 percent
Guatemala:	8 percent
Vietnam:	8 percent
Mexico:	7 percent
Indonesia:	6 percent
Peru:	5 percent
Costa Rica:	4 percent

Nicaragua:	3 percent
El Salvador:	2 percent

That adds up to 90 percent; the rest of the blend would consist of 5 percent coffee from nongrowing broker nation Germany, and fractions of a percent from other producing countries.[11]

How would such a well-traveled global blend taste if pulled from our shelves and shops and concocted together? It would be like throwing every sort of wine, cheap and expensive, red and white, sweet and dry, together in one glass, or like mixing all the colors in a watercolor set together on your palette. The result for wine, paint, and coffee would all be remarkably similar: an awful, muddy brown mess.

Jay Isais's day begins at 6:15 a.m. with—no surprise here—a pot of coffee to share with his wife, Connie. They live near the roasting plant in Camarillo, a town about seventy-five miles north of the Port of Los Angeles. It's a location chosen less out of strategic need and more out of convenience for the founder of the business, who decided to move his home to the picturesque coastal farming area back in the eighties. The Coffee Bean & Tea Leaf's headquarters and distribution center are still in the City of Los Angeles.

Isais's home coffeemaker is an ordinary drip model, his main requirement being a thermal carafe that keeps the brew warm without a heating element underneath. Prolonged heat burns the coffee and ruins the taste, he advises. He is most fond of the African coffees and the Yirgacheffe in particular, but he'll save those for later in the day at work. His wife takes cream, which clashes with the higher acidity of many African coffees, leaving a sour aftertaste. So at home he sticks mostly with dairy-friendly Latin American beans.

At the plant, after his routine housekeeping tasks of reading e-mails and checking the coffee market prices, Isais joins master roaster Jesse Martinez in the cupping lab. Samples of new lots of coffee are air-shipped to the plant on a daily basis, flown in by UPS or FedEx from agents, brokers, and growers all over the world. Martinez has already used a tabletop roaster to prepare today's samples and brewed them up, and an array of small coffee cups are laid out in a circle on the high, round tasting table. Each sample—today they're from Costa Rica, Colombia, and Malaysia—is represented by five cups brewed separately, the better to detect inconsistencies in the sample. Small trays of the beans from each lot are on the table, labeled and displayed to show the appearance of the beans and the roast. The beans are judged by look as well as taste; if there are too many discolored, damaged, or undersized beans, the lot will be rejected on visual quality alone.

The two men begin the cupping when the coffee reaches room temperature, the ideal state for judging. Drinking coffee too hot or too cold diminishes the ability to taste accurately. The two men move around the table, leaning in, sniffing with noses close to each cup, then scooping a bit with a teaspoon. The sound of slurping fills the room. It's quite loud, almost comically exaggerated, but for a purpose: a small amount of coffee is sucked in with a large amount of air in order to disperse the drink across the palate. Then the spoon is dipped in a rinse and it's on to the next cup, the slurps coming in rapid fire, cup to cup, lot to lot, delicate flavors first, then moving to the stronger and more intense brews. The two men move around the table across from one another, circling like prizefighters in the ring.

Afterward, they sit back and grade the coffees on aroma, flavor, sweetness (coffee beans are packed with sucrose), acidity, body, balance, uniformity, and aftertaste, as well as whether the brew leaves a clean cup (as opposed to a gritty one). The last score is the taster's overall impression.

Each category can score a maximum of 10 points, and a total of 80 points is the least a coffee can receive and still be considered "specialty coffee." Isais usually looks for higher scores. Naturally, there's a smartphone app for this: "Cupping Lab," which Isais uses to record the scores on his Android phone. This day, the coffees are all judged well above 80, and orders are placed for all of them.

Meanwhile, on the plant floor, where Martinez had fired up the two roasters for preheating at 3:30 a.m., the roasting is well under way. Five- and six-hundred-pound batches are being roasted in the big imported German Probats, each of them bought for a million dollars. The roasters work like big clothes dryers: a baffled cylindrical drum rotating, heated beneath by gas flames, with one end open to suck in air. When the roasting is complete and the heat turned off, the drum continues spinning and three gallons of water are poured in to cool the beans down and prevent overcooking. The water evaporates almost immediately but it knocks the temperature down in the process.

At the same time, two Italian-made robotic packagers are churning out sealed packages of beans roasted earlier that morning, the machine a hissing whirl of pneumatic moving arms, funnels, and blades. Elsewhere in the plant, two Japanese tea robots are churning out elegant nylon mesh tea bags at a rate of 5,000 an hour in a hypnotic, graceful, and aromatic twirl.

The fresh coffee is packaged in a layered composite material that has a one-way valve in one side concealed beneath the company logo. The valve allows the carbon dioxide from the fresh beans to escape, thereby overcoming the problem of canning and stale coffee. Various sizes of bags are flowing out of the machines: one-pound bags for retail sales, three-pound bags for sale at big-box stores, five-pounders in plain silver bags for making coffee by the cup in company coffee shops.

Before each bag is sealed, oxygen is flushed out with pure nitrogen so the coffee cannot oxidize and spoil inside the bag. In this way, roasted coffee can be kept and retain most of its flavor for months. This is a compromise, as coffee is at its flavorful best twenty-four hours after roasting, Isais says. And yes, he admits, he can tell the difference. But it's still a vast improvement over the old industrial canning process.

The logistics for the Coffee Bean & Tea Leaf are complex: shipments take six to eight weeks to arrive via container from Africa, Indonesia, Central and South America, and Mexico. Two-thirds of the coffee shipments enter the country through the Port of Oakland, which has a preferred rate for certain commodities, coffee among them, and one-third arrives through the Port of Los Angeles. A contract trucking and logistics firm brings the raw green beans, bulk teas and botanicals, and other ingredients to the company distribution center in Los Angeles, where the Coffee Bean & Tea Leaf–branded napkins, cups, lids, and all the other paraphernalia of the modern coffee shop are amassed as well, sourced globally and sold to franchise holders along with the coffee and tea. Once a day, five days a week, a semitruck brings a trailer-load of raw beans, tea, and other materials to the Camarillo plant, drops off the trailer at the loading dock, and takes out a different trailer that has been loaded with the previous day's roasted coffee, bagged and boxed teas, and other finished coffee products. This truck goes back to the LA warehouse and the process repeats the next day.

From LA, short-haul trucks take supplies out for delivery to regional Coffee Bean & Tea Leaf company stores and franchises, nearly half of which are in Southern California. Long-haul trucks depart for more distant distribution points for stores across the nation, and to the port for international franchises and customers. Many of those coffees are returning to areas close to where they were grown, coming full circle over tens of thousands of miles, with Coffee Bean & Tea Leaf outposts in such far-flung locations as Qatar, Saudi Arabia, Vietnam, Bahrain, Germany, Indonesia, South Korea, Mexico, and Mongolia. Espresso-based lattes and mocha concoctions are the big sellers abroad, even more so than in the U.S., as the coffeehouse has become a social magnet worldwide, particularly for the young.

All this is empowered by modern technology and logistics that can ship coffee beans from a distant country to California for roasting, preparation, and packaging, then send them back to the same part of the world for sale as finished product. This transportation-immersed meandering somehow makes economic sense—the magic and the curse of our door-to-door system—and in the process, provides jobs for an estimated 100 million people worldwide engaged in some aspect of the ever-growing coffee business.

The irony isn't lost on Isais that technology almost killed coffee and now is saving it and helping it spread. And yet coffee has in some ways come full circle, he says, returning to that purer experience that hooked all those Civil War veterans so long ago. He points to the training store attached to the roasting plant, a full-fledged coffeehouse to train baristas, open to the public if they can find the obscure location, perhaps by following the aroma that not even the emissions-control afterburners can completely purge. Here you can have that Civil War–era experience of drinking coffee brewed from newly roasted beans, with the scent of roasting still in the air.

"That's the taste that started it all," he says. "Without the battlefields."

Chapter 4

FOUR AIRLINERS A WEEK

The last mile in the Coffee Bean & Tea Leaf supply chain for me—a literal mile—is my short walk to Main Street, where I can find the Ethiopian Yirgacheffe beans that both Isais and I favor.

It's a quick cup for me, and then I have to walk to the bus stop, where one of Long Beach Transit's big red hybrid buses will take me (after one transfer) to San Pedro and a conference on the Port of Los Angeles. The agenda for the conference is all about shipping tech of the future, although the buzz will be all about the crippling port congestion of the present. Busing will take a half hour or so longer than driving my own car, but I won't have to worry about parking, and I can catch up on some reading en route. A reasonable trade-off, with the added bonus of getting a bit of exercise from walking two miles to the bus stop and taking another little notch off the old carbon footprint. The typical household in the U.S. generates the equivalent of 48.5 tons of carbon emissions a year,[1] which is roughly five times the global average—with the single largest chunk coming from transportation.

Figuring out how to get around Southern California by public transit used to be an arduous, frustrating experience—one I'd rarely consider and often regret when I did. Many transit agencies' online guides are atrocious or fragmented. But today's smartphone mapping apps provide a Rosetta stone for bus, trolley, and train travel in many cities in the country with real-time directions, pickup and arrival times, and no peering at incomprehensible bus-stop schedules that make the Zimmerman Telegram decoding project seem like child's play. This may be the single most useful and life-changing capability of the smartphone, enabling every flavor of mobility imaginable, from Waze's traffic jam avoidance to a cornucopia of ridesharing and car lending, to a wave of new personal courier and delivery services, to walking and biking routes that can be very different and significantly better than driving directions. One out of four smartphone owners uses the device to get public transit information, and one out of ten does so frequently.[2] This may help explain why mass transit ridership was higher in 2014 than in the previous fifty-eight years (though, admittedly, Americans' use of transit is far, far lower today than it was a century before the age of the smartphone).[3]

Most of my walk to the bus stop is northbound along the Pacific Coast Highway, which sounds more picturesque than it is. Many parts of State Route 1—PCH to locals—offer glorious views of cliff and coast up and down the state, but my urban stretch is too far inland to provide even a glimpse of the ocean. It's a congested four-lane highway here, and my route takes me across the concrete-lined San Gabriel River that divides Los Angeles County from Orange County and serves as a transition between a mostly residential zone to two big retail, restaurant, and entertainment complexes.

For no good reason, the sidewalks vanish for a quarter mile south of the bridge, forcing pedestrians into the bike lane or the weeds that surround a barren stretch of land where only oil rigs bob like giant, creaky birds pecking for seed. This is the casual way walking is discouraged outside city centers: through thoughtlessness rather than intention, the product of a hundred years of car-centric planning. Thousands of people move from homes in Seal Beach to restaurants, movies, and stores just across the river in Long Beach, but almost nobody walks the short distance because it is so deliberately unwelcoming. It makes for a tense journey as cars whiz by at 50 miles per hour or more, notwithstanding the posted speed limit of 40 miles per hour. The sidewalks reappear on the bridge itself and continue on the north side, but then the bike lane vanishes without warning a bit farther up the highway, turning into a third lane of car traffic. Cyclists who don't know to avoid this trap suddenly find themselves in the mix with speeding cars about to turn into their paths, a potentially deadly defect in detail and design that is also all too common throughout the region (and much of the country).

Today I see a bike rider miss serious injury by inches as I walk along this area. The car that nearly hit him while making an ill-advised lane change does not fare so well, barely avoiding the bicyclist but then rear-ending an SUV with a loud crunch. Traffic is blocked as the drivers stop and get out to assess the minor damage and to make the ritual exchange of shrugs, grimaces, accusing stares, and insurance information.

This is a mundane, everyday event, the sort that city dwellers barely notice anymore, like the near misses pedestrians regularly experience in crosswalks. But this one I do remember because of what the bicyclist says to me. I was walking right by him and so offered a commiserating shake of the head at his close call, at the obliviousness of some drivers, and at this zero-sum game of an intersection. He had lugged his bike out of the street onto the sidewalk and was standing astride it, running a hand through his long hair, calming himself. "That's the third time this has happened to me," he tells me in a shaky voice.

"Really?" I ask. "You mean this week?"

He shakes his head. "Today."

The car is the star. That's been true for well over a century, an unrivaled staying power for an industrial-age, pistons-and-brute-force machine in an era so dominated by silicon and software. Cars conquered the daily culture of American life back when spats, top hats, and child labor were in vogue, and well ahead of such other game changers as radio, plastic, refrigerators, the electrical grid, and votes for women.

Cars—all 1.2 billion of them worldwide[4]—may not be the most vital component of our sprawling transportation landscape, or the most economically potent; the goods movement fleets and flotillas hold those crowns. Our beloved internal-combustion–powered, 3,977-pound metal boxes on wheels aren't even the most irreplaceable slice of the transportation pie no matter how attached we are to them or helpless we feel without them.[5] They are still very much the same beast as Henry Ford's Model T, refined, safer, improved (though not as much as you think), yet still of the same basic, terribly inefficient DNA. Our cars are so rooted in that past they have never shed their deep connections to the age of horse transport, with the car's shape and dimensions still based on horse-drawn carriages and engine output still measured in the archaic eighteenth-century metric of horsepower. Horsepower! Do we measure power plants and nuclear reactors and computer processors by how many horses they "equal"? The surprise, from a technological perspective, is that the conventional car wasn't replaced by any number of more modern designs or technologies long ago, just as cell phones eclipsed landlines and then smartphones dethroned dumb ones.[6] The same story holds for coal-fueled cargo ships, steam locomotives, dirigibles, telegraphs, phonographs, typewriters, vacuum tubes, and film cameras.

And yet the car remains the star. It's how generations of Americans have experienced transportation. It's how we intuitively measure distance, not in terms of miles but in car time: *Oh, that store is just fifteen minutes away.* Cars are intrinsic to our culture. We associate them with personal freedom, and we incorporate them into all our big moments from the start of life to the end. We gaudily decorate our cars for the post-wedding cruise into married life. We drive our newborns home from the hospital in a flurry of photos and Facebook posts, and gift expectant mothers with car seats at their baby showers in preparation for that first ride. We go to work in them, we take our meals in them (19 percent of them nationwide, by some estimates),[7] we date and mate in them. We lavish them with polishes, waxes, personal decor, religious symbols, and political slogans, then show them off like prized Thoroughbreds. And at the end, we have built fleets of lustrous black cars for our last rides to the cemetery.

The car's outsized footprint even governs where and how we live. America has organized its built landscapes around cars to enable their movement, their parking, their convenience, and our dependence on them. The country has enough parking spaces to cover every inch of Delaware and Rhode Island combined—as many as eight spaces for every car in the country, which adds up to about 30 percent of open space in the dense cores of our cities.[8] Our emotional involvement with our cars is no less outsized. We spend billions on new lanes and high-tech traffic control centers just in the *hope* of shaving a few minutes off our travels—billions for mere minutes—because movement impeded, even for the meager interval we'd happily invest in awaiting a restaurant table or a beer at the ball game, is psychologically unbearable if it takes place in a car. Researchers have documented this phenomenon time and again: the brain perceives each minute of a travel delay—waiting for a bus, looking for parking, being stuck in traffic—as two to three times longer than a minute spent moving freely. Humans are conditioned, perhaps even hardwired, this way. This may explain why voters so often prefer spending on such projects as Carmageddon and their often fruitless promise of faster travel rather than investing in buses and subways that, by definition, may be more resource-efficient but also involve that hated, psychically torturous wait time caused by schedules, stops, and stations.

Before cars, streetscapes were designed primarily for walking. The rules of the road were simple: pedestrians ruled. When streetcars came along, they were open: passengers could just jump on as they passed. Now most thoroughfares are designed for driving, and if anyone is jumping, it's to jump out of the way. In many locations, sidewalks are either forbidden (walking on freeways is illegal in most states except in emergencies, and death-defying even then), omitted (as a cost savings), or in disrepair (Los Angeles's notorious city sidewalks being exhibit A, a literal walk of shame as even the mayor has conceded).[9] In some states, pedestrians are legally required to avoid conflicts with cars rather than the reverse, and laws and custom so favor automobiles that driving into someone on foot is often treated as a traffic offense rather than an assault, if a citation is issued at all—even in cases of pedestrian injury or death.

Finally, while it is true that the Millennial generation is trending away from driving and owning cars,[10] and a scattering of cities are building robust bike, transit, and walking-friendly zones for a more equal sharing of the road, this is a slow and controversial trend, still too nascent to undermine car culture primacy. For sixteen-year-olds all over the country, surviving the ordeal of the driving test and the triumphant receipt of that laminated icon remains America's one great secular ritual and rite of passage, as resonant in its own way as baptism and bar mitzvah, confirmation and vision quest. The license marks an end to childhood, a new independence, and a kind of machine-powered freedom in which the car is not just conveyance but emblem—the star of the show.

All of this history, culture, ritual, and man-machine affection helps explain why the true cost and nature of cars have become so very hard for us to see. And what is that nature? Simply this: in almost every way imaginable, the car, as it is deployed and used today, is insane. And not in a good way. More like the deep-fried Twinkies stuffed with caviar I saw being sold for $125 apiece at the county fair this summer—insane that way. Except our cars are much more likely to kill us.

But wait. Cars are the epitome of convenience, aren't they? That's the allure and the promise that's kept us hooked, dating all the way back to the versatile, do-everything, on- and off-road miracle of the Ford Model T. Convenience—some might call it freedom—is not a selling point to be easily dismissed—this trusty conveyance, always there, always ready, on no schedule but its owner's schedule. Buses can't do that. Trains can't do that. Even Uber makes you wait. Whenever a car owner wants to jump in and go, the car is there waiting. No sweating and pedaling. No hours spent walking. This is how America gets to the store. This is how America gets to work. This is how pizzas come to our door. The car's not insane, it's amazing, right?

But there's a catch. The price for this convenience is acceptance of vehicles that are nothing less than rolling disasters in terms of economics, environment, energy, efficiency, climate, health, and safety. Our failure to acknowledge the social and real-dollar costs of these automotive shortcomings amounts to a massive hidden subsidy. The modern car could not dominate, or exist at all, without this shadow funding.

So what are the failings of our cars? First and foremost, they are profligate wasters of money and fuel: more than 80 cents of every dollar spent on gasoline is squandered by the inherent inefficiencies of the modern internal combustion engine.[11] No part of our infrastructure and daily lives wastes more energy and, by extension, more money than the modern automobile.

While burning through all that fuel, our cars and trucks spew toxins and particulate waste into the atmosphere that induce cancer, lung disease, and asthma. These emissions measurably decrease our longevity—not by a matter of days, but years. The Massachusetts Institute of Technology calculates that 53,000 Americans die prematurely every year from vehicle pollution, losing ten years of life on average that they would have survived in the absence of tailpipe emissions.[12] There are also the indirect environmental, health, and economic costs of extracting, transporting, and refining oil for vehicle fuels, and the immense national security costs and risks of being dependent on foreign-oil imports for significant amounts of that fuel.

As an investment, the car is our most massive waste of opportunity—"the world's most underutilized asset," investment firm Morgan Stanley calls it.[13] That's because the average car sits idle 92 percent of the time. Accounting for all costs, from fuel to insurance to depreciation, the average car owner in the U.S. pays $12,544 a year for a car that puts in a mere fourteen-hour work week. Drive an SUV? Tack on another $1,908.[14]

Then there is the matter of climate. Our movement from door to door is a principal cause of the global climate crisis, exacerbated by our stubborn attachment to archaic, wasteful, and inefficient transportation modes and machines.[15] But are cars the true culprit? What about other modes of travel? Airplanes, for instance, are often singled out as the most carbon-intensive form of travel in terms of emissions per passenger mile (or per ton of cargo). By some estimates, carbon emissions from one round-trip jetliner trip between New York and Los Angeles generates more than 10 percent of the average American's total carbon footprint for an entire year.

But that's not the whole story—a case of statistics masking, rather than illuminating, a larger truth. Total passenger miles by air are miniscule compared to cars: in any given year, 60 percent of American adults never set foot on an airplane, and the vast majority who do fly take only one round trip a year. If air travel were the only source of transportation emissions, our climate future would be bright rather than blighted. Unfortunately, air travel is not our primary problem, contributing only 8 percent of U.S. transportation-related greenhouse gases. Cars and trucks, by contrast, pump out a combined 83 percent of transportation carbon.[16]

Furthermore, airlines—whose margins have been continually squeezed by fuel costs and competition—out of necessity have become far more efficient in terms of the energy expended moving one passenger one mile. Some of these gains came through better engineering, but just as much arose from more efficient booking systems that keep planes profitably (and uncomfortably) full. Overall, airliners have become 74 percent more efficient than they were in 1970—to the point that it is worse for the climate to drive long distances in an average American car than it is to fly. Driving your SUV or even a mid-size car from New York to LA is worse for the planet than flying there.[17] This is true in part because car fuel efficiency has improved far more slowly than planes, but also because of Americans' increasing propensity to drive alone, which has made car travel *less* efficient and more carbon-intensive per passenger mile in recent years.

So cars pose the biggest threat on the climate front, with all the costs that global warming imposes on our infrastructure, homes, and lives through increasingly severe storms, droughts, rising sea-levels, and pressure on food supplies. If the price of gasoline and the vehicles that burn it actually reflected the true costs and damage they inflict, the common car would be as extinct as the dinosaurs. Gasoline would cost way more than $10 a gallon. That's how big our secret subsidy is.

And that's not even counting the most dramatic cost we accept and subsidize: cars waste lives. They are one of America's leading causes of avoidable injury and death, especially among the young.

The materials, techniques, and technology to shrink or fix each of these categories of car shortcomings exist today, and all of them would be cheaper in the long term than allowing the status quo to persist. But the *will* to accept or even see the shortcomings of the car is another story.

Oddly, the most immediately devastating consequence of the modern car—the carnage it leaves in its wake—seems to generate the least public outcry and attention. Jim McNamara, a sergeant with the California Highway Patrol, where officers spend 80 percent of their time responding to car wrecks, believes such public inattention and apathy arise whenever a problem is "massive but diffuse." Whether it's climate change or car crashes, he says, if the problem doesn't show itself in a big bang all at once—as when an airliner goes down with dozens or hundreds of people on board—it's hard to get anyone's attention. Very few people see what he and his colleagues witness daily and up close: what hurtling tons of metal slamming into concrete and brick and trees and one another does to the human body strapped within (or not strapped, all too often). Short answer, McNamara says: nothing good. Nothing you want to see on a full stomach.

In contrast, a typical driver's experience of car violence is radically different, little more than a glance at a wreck on the roadside, some broken glass and bent metal briefly visible in a flash, only to disappear in the rearview mirror a moment later.

Mostly it's the machine that's visible in such moments; any human damage is hard to see, as bloodstains and bodies are quickly draped by innocuous sheets of yellow or orange plastic, while the living are immediately removed for treatment or shelter. What may have caused the wreck, the how and why and who of it, is rarely apparent to passersby. So, more than anything, a roadside wreck is experienced by the vast majority of drivers as a nagging but unavoidable inconvenience—just another source of detours and traffic jams, a bottleneck to be passed with relief. Increasingly popular and powerful smartphone traffic apps eliminate even those brief close encounters with the roadway body count, routing savvy drivers away from crash-related congestion. The typical car wreck is becoming all but invisible to everyone but those who are killed or maimed and those whose job is to clean it up. Many are aware at some level that troubling numbers of people are injured and die in cars, but most remain unfazed by this knowledge.

The contrast couldn't be greater with public perception of airliner crashes, which always generate a high-visibility tsunami of fear, headlines, and spare-no-expense investigations. As counterintuitive as it may seem when comparing passenger-laden airliners with the crash of one car carrying one person, this disparity in attention cannot be justified by the numbers. Quite the contrary: in the fourteen years following the terrorist attacks of 9/11, there were eight crashes on American soil of passenger planes operated by regional, national, or international carriers. The death toll in those crashes totaled 442. That averages out to fewer than three fatalities a month.[18]

The death toll on America's streets and highways during that same period since 9/11 was more than 400,000 men, women, and children.[19] The traffic death toll in 2015 exceeded 3,000 a month.

When it comes to the number of people who die in car wrecks, America experiences the equivalent of four airliner crashes every *week.*

The response to that automotive death toll is telling. In the twelve months between June 2014 and May 2015, Americans took 779 million plane trips. That's a big number, and air travel is a vital part of the mobility picture; but when it comes to how Americans move, it's barely a rounding error. Americans take 1.1 billion trips in cars every *day.*[20]

When it comes to car crashes, there are none of the nationwide safety bulletins or mandatory pilot training that so often result from aviation crash probes. No sweeping investigations of a single wreck nor attempts to find ways to prevent similar tragedies in the future, as is compulsively done in the aviation arena. There are just routine reports by local coroners and police officers providing bare details of what happened, and not a word of guidance on how to avoid the next one. Most crashes—including fatal ones—are not even reported in the news media. Those that do make the news just describe the damage to person and property but almost never report the cause, if one is ever formally determined. There are, of course, lawsuits, America's go-to response to anything that causes injury or death, but almost all of them are resolved quietly by insurance companies. The only exception is when they involve safety defects in the machine, not the driver, in which case there is often intense media coverage and public concern, although in the fifty years the feds have had the power to recall cars for safety reasons, such defects have caused only a tiny fraction of the deaths and injuries experienced on the highways every day.

A normal day on the road, then, is a "quiet catastrophe," as Ken Kolosh, the statistics chief for the National Safety Council, calls it. He ought to know: he makes his living crafting the annual statistical compendium of every unintentional injury and death in the country. Kolosh is America's amiable and understated Dr. Death, poring over those coroners' findings, police reports, and disparate databases from state and federal agencies to construct a picture of how we kill ourselves. It's his job to spot the trends, to note such curiosities as the accelerating number of accidental poisonings (mishandled prescription drugs, mostly) and the fact that the battle to reduce drunken driving has stalled for decades. And, quiet or not, Kolosh says, the numbers that cross his desk show that, despite gradual declines, the automotive casualty count remains nothing short of catastrophic.

Car crashes are the leading cause of death for Americans between the ages of one and thirty-nine. They rank in the top five killers for Americans sixty-five and under (behind cancer, heart disease, accidental poisoning, and suicide).[21]

One out of every 112 Americans is likely to die in a traffic crash.[22] Just under 1 percent of us.

The economic costs and societal impact from motor vehicle death and injury each year amounts to $836 billion. The direct economic costs alone—the medical bills and emergency responder costs reflected in all our taxes and insurance payments—represent a tax of $784 on every man, woman, and child living in the U.S.[23]

The numbers are so huge they are not easily grasped, and so are perhaps best understood by a simple comparison: if our roads were a war zone, they would be the most dangerous battlefield the American military has ever encountered.

First, there is the annual death toll from motor vehicle crashes in the U.S.: 35,400 dead in 2014, by Kolosh's count. That number is greater than the annual U.S. military death toll during each war America has ever fought except the Civil War and the two world wars.[24] That means U.S. highway fatalities outnumber the yearly war dead during Vietnam, Korea, Iraq, Afghanistan, the War of 1812, and the American Revolution.

Now consider the annual U.S. injury toll from car wrecks—just the ones serious enough to require emergency room trauma care: 2.5 million.[25] Those wounded on the highways exceed the numbers of wounded *and* dead in World War II, World War I, Vietnam, Korea, Afghanistan, Iraq, the War of 1812, and the Revolution *combined.* That's not just the yearly casualties from those wars. One year of car crash injuries and deaths in the U.S. is greater than all the dead and wounded from the *entire duration* of all those wars combined, with numbers to spare to cover all Union Army dead and wounded from the Civil War as well.

Widen the category to include a year's worth of U.S. motor vehicle injuries that required some medical consultation, and those 4.3 million injured[26] far outnumber the military dead and wounded in every war and conflict in which America has ever participated, Confederate casualties included.[27]

One year of driving in America is more dangerous than all those wars put together. The car is the star.

Chapter 5

FRIDAY THE THIRTEENTH

Friday, Febuary 13, 2015, was a normal day on America's roads, all 4,071,000 miles of them,[1] a web of asphalt, concrete, and unpaved dirt that binds our nation and neighborhoods, linking Wall Street to Main Street, port to warehouse, shopper to store—and driver to doom.

A young entertainer walked home rather than drive drunk, only to be run down by a drunken driver. A graphic artist drove oh-so-carefully in the snowy darkness, only to collide head-on with a car barreling down the wrong side of the interstate. And there was the crazy mundanity of a mattress dropped on a freeway, a soft object, a thing that cushions—except when a car strikes it at high speed. Then it sets off a fatal chain reaction, a deadly game of pinball in which the balls weigh two tons and the bumpers are concrete medians.

America's roads that day, like every day, bore casualties around the clock:[2]

A death every 15 minutes.

A trip to the emergency room every 12.6 seconds.

An injury serious enough for a medical consult every 7.3 seconds.

And there was a crash of some kind, somewhere—involving

death, injury, or property damage, or a mix of all three—every 2.8 seconds. It's happening right now: in the time it takes to read this sentence, two more car crashes occurred on America's roads and streets.[3]

The day began with a wreck in New Hampshire so bad, it closed a frigid highway in both directions: a wooded stretch of Route 16 in the town of Milton, just short of the Maine state line. The road had to be scrubbed of cars and trucks so the first responders could descend on the twisted wreckage, trying to stanch the blood, then extract the occupants and attempt to resuscitate those who showed no signs of life but who might yet be saved. Next came the investigators, followed by the coroners, then the tow trucks and, finally, the cleanup crews to carry off the post-crash debris and the detour signs, letting the inexorable flow of traffic resume. Four hours later, a bit past midnight, no sign remained of the three women and the family dog who perished in a head-on collision. The crash occurred while it was still Thursday, but its aftermath spilled over into Friday the thirteenth.

Most days in America begin with a lane or street or highway closed somewhere because of a fatal crash the night before.

Victoria Rose had been driving south on the two-lane highway toward her home in Revere, Massachusetts, when her 2005 Jeep Liberty SUV drifted across the centerline and into northbound traffic. This would be the first of many such "over-the-line" wrecks this day, a common sort of accident. The fifty-seven-year-old died instantly when she collided head-on with a Subaru Outback station wagon that had been cruising quite legally and properly in the northbound lane. That driver, Allison Smith, an environmental expert with the nonprofit New England States Committee on Electricity, survived only long enough to be cut free from her mangled car and rushed to a local hospital, where the thirty-one-year-old died within minutes. One of her passengers, Vanessa Cox of Boston, athletic department administrator at Brandeis University, also died in the hospital emergency room. A third person in the Subaru, Lucy Pollard, a prep school teacher and Allison Smith's wife, survived with serious injuries, as did a passenger in Rose's Jeep. A small dog riding in that car had to be euthanized at the scene.

Wandering over lane lines, either off the road to the right or into oncoming traffic to the left—"road departure" and "lane departure" are the official terms—can be a hallmark of distracted driving or dozing at the wheel. But, as is often the case, establishing with certainty the cause for Rose's fatal swerve has been difficult for investigators. Those most able to explain were killed.

Friday the thirteenth was full of such unexplained drifts into destruction. Jillian L. "Jilly" Rebel was a victim of one. A forty-year-old worker at a local gas station and convenience store, Jilly was well-known in her community for her cheery greetings to customers and her love of all sorts of animals. On Friday the thirteenth she drifted across the centerline on I–80 in Lackawannock Township, Pennsylvania, at 2:30 in the afternoon. Her PT Cruiser hit the grassy median, skidded, and rolled four times before crunching to a halt. With her seat belt left unbuckled as she drove—another common theme this day—she hurtled out of the car to her death during the roll.

Ten minutes later, eighty-seven-year-old Dick Morgan died at the wheel of his Ford F–150 pickup on Minnesota State Highway 19 after he, too, drifted across the centerline and collided head-on with a semitruck. Chad Hilborn, a thirty-three-year-old corrections officer coming home from a graveyard shift in Washtenaw County, Michigan, died the same way, crossing the centerline and running head-on into an approaching pickup truck, critically injuring the other driver. There were more than two dozen fatal crashes of this sort during the day: drifting out of lanes, drifting into clearly marked barriers, drifting into fields and forests. None of these involved mechanical failures or adverse weather, nor was there evidence of drunken driving, which means avoidable errors—distraction, dozing, or speeding—were involved.

Extensive research in the U.S. and United Kingdom dating back to 1979 pins the blame for 90 to 99 percent of traffic crashes on human error.[4] Speeding (a factor in 30 percent of traffic deaths) and distraction (26 percent) together account for more than half of all crashes.[5] The single most frequent cause of fatal collisions, driving drunk (30.8 percent of road deaths) is certainly an example of human error, as are all the other acts of reckless, negligent, incompetent, and criminal driving behavior: tailgating, running red lights, refusing to yield to pedestrians, ignoring the right of way of bike riders, driving at unsafe speeds, driving drowsy (instead of pulling over when drowsy). All of these are intentional behaviors, as opposed to the commonly used but almost always incorrect descriptor "accidental." The fatal results might have been unintended, but the behavior is no accident.

Driving is by far the most difficult, complex, and high-risk task most people other than bomb defusers and brain surgeons will ever do. Yet driver training and licensing tests set the bar so low that almost everyone passes (eventually). So it's not really surprising that error, by commission or omission, is the primary culprit in almost all car wrecks. The surprise would be if it were otherwise.

The real surprise is that, for all its prevalence, few mechanisms exist to deal effectively with the often reckless or negligent decisions that precede fatally bad driving—either preventing it, minimizing it, or imposing consequences to deter it.

On the afternoon of Friday the thirteenth, Tiffanie Strasser waited patiently for the "Walk" signal and a green light before stepping out to cross a busy street in Denver's lively Bonnie Brae neighborhood, pushing her two children before her in a double stroller. There was five-year-old Audrey, blind and developmentally delayed, and, next to her, three-year-old Austin, Audrey's precocious, self-appointed "big" brother, protector, and guide, who was counting down the days to his fourth birthday one week away. The family was headed for ice cream, then the library.

At the same intersection, also with a green light, was driver Joan Hinkemeyer, a seventy-eight-year-old retired professor and current gardening columnist for the *Washington Park Profile*, a monthly community newspaper. Hinkemeyer, after departing from her volunteer work at the same neighborhood library the Strassers had planned to visit, turned her car left—into the crosswalk, and into the Strassers. Tiffanie and Audrey suffered only minor injuries, but Austin was dragged several feet and suffered fatal head trauma. Hinkemeyer said she couldn't see the family in the crosswalk because of glare from the late-afternoon sun.

After a long investigation, the Denver district attorney charged her with one count of careless driving resulting in death, and two counts of careless driving resulting in injury—misdemeanors under Colorado law, punishable by a maximum fine of $1,000 and a year in jail. The maximums were only theoretical, however; lesser penalties are the norm. Prosecutors offered a plea bargain with a sentence of no jail time, two hundred hours of community service, and a driving class in exchange for an admission of one count of careless driving.

In an emotional court hearing in which Hinkemeyer apologized to the Strasser family, Austin's parents pleaded with the judge to reject the deal in favor of a maximum jail sentence and a ban on Hinkemeyer ever driving again. After setting up framed photos of her son throughout the courtroom, Tiffanie Strasser stared at the judge and asked, "How can the consequences of killing a child in a crosswalk be some hours of community service?"

The judge did indeed reject the plea bargain but accepted a slightly modified deal, hastily worked out on the spot: one day in jail plus thirty days of home confinement and a requirement that Hinkemeyer pass a driving course in order to keep her license.

These would be the toughest sanctions imposed this day anywhere in the nation for fatal driving, other than cases involving intoxication. A crash earlier that day in Auburn Hills, Michigan, was far more typical. A thirty-three-year-old machinist, apparently in reaction to erratic driving by another car southbound on Interstate 75, lost control of his Ford Escort, struck the concrete median wall, and came to a stop in the left lane. The disabled car was then struck by a Chevy Malibu. The driver of the Ford, John Steele, who had recently purchased a house he was immersed in renovating, suffered critical injuries and died three weeks later from complications from cerebral trauma. Eyewitnesses thought a black Ford SUV had clipped Steele's car and fled after causing the deadly sequence of events. After police issued a plea through the media for additional witnesses, the driver of the mysterious SUV came forward. Investigators determined that vehicle had no physical contact with Steele's car after all—it was just a close call. The SUV driver's consequence: a ticket for erratic driving.

Even more typical: a driver in Conroe, Texas, at 6:30 that morning turned left in front of a young motorcyclist going 55 miles an hour in the opposite direction on busy State Highway 105. There was no way for the biker to avoid slamming into the car that suddenly veered into his path. Seventeen-year-old Trenton Fortune, on his way to meet his girlfriend for breakfast before they went to their high school together, died from the impact. The driver of the car was blamed for the crash in police reports for failing to yield the right of way, but no criminal charges were filed.

There's an old joke told around the California Highway Patrol: *If you want to get away with killing someone, use a car.*

It's not a very funny joke—just a literally true one. As long as a driver is sober, causing a crash that leads to injury or death is rarely treated as a crime, and rarely leads to meaningful sanctions, such as taking away someone's driving privileges (although such a policy might have limited impact, as a 2008 study found: one in five fatal collisions involve a driver with no valid license).[6] Erratic driving may cause a chain reaction leading to hospitals and graves, but such arguably negligent or reckless conduct is almost always regarded as an ordinary traffic violation or a simple "accident." On the rare occasions when prosecutors file formal criminal charges such as vehicular homicide, they often end up reduced to misdemeanors or traffic violations. Case in point: a twenty-year-old North Dakota woman, accused in 2014 of using her phone while driving, was allegedly so inattentive to the road that she rear-ended another car at 80 miles per hour without ever braking. An eighty-nine-year-old great-grandmother in the other car died. The case against the driver started as a stiff felony prosecution but ended up plea bargained down to a misdemeanor, with the young woman receiving a year of unsupervised probation and a guarantee that her record would be expunged if she finished the year without incident.[7]

In other cases, the lenient treatment is a function of a system deliberately constructed to excuse auto violence, as in the 2013 death in New York City of three-year-old Allison Liao, run down as she walked hand in hand with her grandmother in a crosswalk with the right of way on Main Street in Flushing. The Queens district attorney refused to prosecute the driver of the SUV that killed the toddler because he wasn't drunk or on drugs at the time; the driver was given a pair of traffic tickets instead. And even those were tossed out by a callous and disinterested judge—abetted by an inept department of motor vehicles—who conducted a forty-seven-second hearing without even reviewing the facts of the case. The system, designed to process as many tickets in a day as possible, had neither time nor concern for the death of a three-year-old.[8]

Such outcomes are the rule, not the exception, although in most cases stiff initial changes are rarely filed in the first place. In 2014, the *Wall Street Journal* combed through traffic and criminal justice data to show that 95 percent of traffic fatalities in New York City led to no criminal prosecutions.[9] A similar analysis in Oregon showed that sober drivers in fatal car crashes in the Beaver State had little to worry about, either.[10] Drivers, it seems, enjoy an exalted position in a U.S. criminal justice system that is otherwise viewed as one of the toughest in the developed world.[11]

No other single product has shaped the law, the landscape, and public opinion as radically as the car. As revolutionary as the horseless carriage was at redefining travel in America, other changes instigated by cars went deeper, none more than the revealing evolution of how we describe car violence. A common term of the 1920s, "motor killings," has morphed into today's "fatal accidents." The gulf between the two ways of describing road death—the early term, angry and accusatory, and the current phrase, blameless and bureaucratic—perhaps explains what might otherwise seem inexplicable: the American public's willing acceptance for decades of cars designed and used in such a way that they have become the number one killer of our children and teenagers and everyone else under forty.

Sheriff's investigators combing the wreckage on the night of February thirteenth in Forsyth County, Georgia, concluded high school senior Taylor Oliver was using his cell phone to send text messages in the moments before his Ford pickup truck drifted off a county road. When he realized his blunder, Oliver appeared to try to steer back onto the asphalt, but he overcorrected and lost control. His pickup slid sideways across the centerline and into opposing traffic. A "Super Duty" Ford F–350 pickup truck—towing a twenty-four-foot trailer with a Bobcat front-end loader onboard—broadsided the passenger side of Oliver's pickup. Both trucks careened off the roadway and down an embankment. The driver and passenger in the bigger truck were not injured, but Oliver, alone in his Ford, suffered fatal head injuries.

Investigators in Greensboro, North Carolina, theorized that some sort of distraction also sent UPS Store worker Roger McHenry to his death when his car careened into the "gore point" where Interstate 40 splits, and in Flagstaff, Arizona, where the driver of a semitruck hauling a load of beer died when he drove off the road and into a culvert on Interstate 40, closing the westbound side of the freeway for half a day because of spilled beer and debris. And Mandy J. Theurer died on the familiar rural road where she lived, another likely distracted driver, careening into a ditch she had driven by safely countless times before. The twenty-six-year-old cosmetologist who worked at the Cute as a Button salon in Portland, Indiana, did not have her seat belt buckled, and when her car overturned she was ejected, suffering fatal trauma. Fifty percent of all car crashes occur within five miles of home.

Because of the prevalence of such accidents, distracted driving—particularly when it involves using cell phones—has in recent years become the focus of public debate, legislation, police crackdowns . . . and considerable misunderstanding.

The problem is not cell phones per se—it's the human brain and what the National Safety Council calls "the myth of multitasking."[12] This is the common but incorrect belief that humans are good at doing several things at once, whether driving or cooking or dancing. In most endeavors, this is a harmless belief; but when traveling 65 miles per hour, the stakes are too high to be governed by myth. We're not talking about walking and chewing gum here, but about tasks that require thought and attention. The truth is, no matter how it feels or appears subjectively, the human brain cannot pay attention to two cognitive tasks at the same time. Computers can do this but not humans. We are not built for it.

What the brain is really good at is *toggling* between mind-absorbing tasks—shifting focus rather than dividing it, then picking up where it left off when it toggles back. So when drivers are messing with cell phones or car stereos or dropped baby bottles, they are not driving. They have toggled, shifting focus and attention from one task to another, sometimes quite rapidly, but never simultaneously.

This is the essence of distraction and it's not limited to staring down at a phone instead of out through the windshield. Brain scans of drivers talking on the phone while staring straight ahead show that activity in the area of the brain that processes moving images decreases by one third or more—hard evidence of a distracted brain. There have been many fatal crashes attributed to this "inattention blindness," commonly called "tunnel vision." Drivers talking on cell phones or performing other non-driving tasks can become so focused on the non-driving activity that their brains fail to perceive half the information their eyeballs are receiving from the driving environment. They can appear to be paying attention—the drivers may even *think* they are paying attention—but they are distracted drivers. This is not a matter of skill or practice or experience. It's biology.

The U.S. Department of Transportation made a public service video a few years ago featuring a fatal crash in Grand Rapids, Michigan, that illustrates this problem.[13] A twenty-year-old woman, while driving on a city street, spoke by cell phone with someone at the church where she did volunteer work with elementary school–aged children. Witnesses later told police that she was looking straight ahead out the windshield while talking on her phone—not looking down, not texting, not dialing. Yet she drove right through a red light at an intersection into cross traffic. The light had been changed long enough for several cars to move through the intersection right in her field of view, yet she kept going, never braking but speeding into the intersection against the light, where she broadsided another car at 48 miles an hour. A twelve-year-old boy in the car she broadsided was killed in the collision.

The problem isn't that humans can't be good drivers. When focused and fully engaged, humans can be fantastic drivers, capable of displaying the same sort of innate mental calculation of trajectories and safe passage as a star football quarterback and his receiver use to complete a pass forty yards downfield. At the same time, the brain is very good at ignoring false signs of danger—thereby avoiding the constant traffic slowdowns that would result from stopping for every possible hazard—and able to distinguish a person running into traffic from a wayward balloon blowing into the road without a moment's hesitation. That might sound like a no-brainer, but that balloon would give fits to the literal computer brain of a driverless car, which would likely grind to a halt for the balloon as quickly as for a person.

The problem is that staying focused and fully engaged for sustained periods of time is not what humans do best, with or without a cell phone in hand. Driving can be boring, the conditions monotonous, or the street or intersection so familiar that we hardly notice or think about it. That's when the brain's ability to stay focused on safe driving decays, because humans are terrible at doing repetitive, routine tasks while also remaining alert for sudden, unexpected dangers that may or not materialize. Robots and computers excel in that scenario—which is why autopilots on airliners have become essential tools during long flights.

Using cell phones, employing GPS devices, and engaging in other tasks unrelated to driving (eating, putting on makeup, using an electric razor—almost anything can become a distraction) heighten this risk. And so distracted driving happens every day in America. The results are fatal: about once an hour that we know of.

When danger looms—a car jamming on the brakes in front, a stop signal unnoticed, a curve in the road when attention is diverted—a lot of bad things can happen before a distracted driver can refocus attention and react.

"You put your life in the hands of *everybody* passing you on a two-lane road every day," Jim McNamara says, and the CHP officer's tone makes clear that he does not enjoy this particular aspect of driving. McNamara knows that trust is violated daily. He's seen it. He's cleaned up after it. It happens so fast and so easily: the physics of four-thousand-pound objects racing toward one another, pitted against the temptation so many drivers indulge, to read that text or find that playlist, because it'll only take a few seconds looking away from the road. The problem: two cars traveling 65 miles per hour in opposite directions have a closing speed of 190 feet per second. All you have to do is look down at your phone and inadvertently drift across the centerline: in just three seconds, two cars that had two football fields' worth of distance between them are in a head-on collision.

"I wish we could impress that on people, the ramifications of looking down at your phone," McNamara says. "We see it all the time."

Even when there is time to react, the sudden refocus from a distracted state can lead to overcorrection—a too-vigorous swerve of the steering wheel or skid-causing jam of the brakes that can make things worse. This, too, happened several times on Friday the thirteenth.

For the police, proving a case of distracted driving is difficult. Unlike drunken driving, which can be verified with blood alcohol testing, or speeding, which can be calculated from traffic cameras, impact force, skid marks, and physical damage, distraction rarely leaves hard evidence and, in the absence of witnesses, can be inferred though seldom proved.[14] But now an unusual and sobering analysis of 1,700 videos of teen drivers taken from in-car recordings of crashes suggests distracted driving may be a much larger problem than previously believed. Two particular types of crashes stood out: 89 percent of road-departure crashes—in which cars drifted onto shoulders or off roadbeds entriely—and 76 percent of rear-end collisions were caused by distraction.

The videos are horrifying, one crash after another in which death or major injury was avoided by luck rather than skill: teens staring down at cell phones for four seconds, applying makeup, staring out the side windows—all while their cars veered off the road, slammed into the back of other cars, or spun out as the drivers looked up and overreacted. The seven most common distractions observed in this video study played out this way:

- Interacting with one or more passengers: 15 percent of crashes
- Cell phone use: 12 percent of crashes
- Looking at something in the vehicle: 10 percent of crashes
- Looking at something outside the vehicle: 9 percent of crashes
- Singing/moving to music: 8 percent of crashes

- Grooming: 6 percent of crashes
- Reaching for an object: 6 percent of crashes

The study, released in March 2015 by the American Automobile Association Foundation for Traffic Safety, found that distraction was a factor in nearly six out of ten moderate to severe crashes involving teen drivers.[15]

Kids who drive to Lincoln-Way Central High School in the Chicago suburb of New Lenox every day would do almost anything to avoid the $125 annual parking fee for a spot in the school lot. The preferred method to beat this bill is to park on the far side of US Highway 30, which the school campus sits astride. The drawback of that free parking is the need to brave a busy four-lane thoroughfare at rush hour in order to get to class. Students can start to cross, then find themselves stranded on the concrete median separating the eastbound and westbound lanes, forced to wait for traffic to ease. So it was with seventeen-year-old Dylan Wischover of Manhattan, Illinois. Whether he tripped, slipped, or simply misjudged when to cross, the junior stepped off the median and into the path of a westbound semitruck. The trucker had no chance to brake before striking and killing Dylan. Grief counselors were summoned to the school to help distraught students, while at home his parents had to face their son's room, filled with boxes of pickup truck parts he and his father had planned to use that weekend in Dylan's favorite pastime: fixing up old vehicles.

Among his final Twitter posts the day before he died, one was wryly humorous, the other darkly poignant: "The only 4.0 I'm getting this semester is when girls rate me out of 10," he wrote, followed by the post, "Ride or die."

The uneasy mix of pedestrians and vehicles played out with

terrible consequences many times on this particular Friday the thirteenth. A seventy-year-old finance professor at the University of Arizona, out for a stroll at 5:00 p.m. along the shoulder of Tucson's aptly named Speedway Boulevard, was struck from behind by a car coming around a bend in the road. Sharon Garrison, whose online tributes included praise from students, colleagues, and business leaders she previously taught during a long career, died the next day of her injuries. The driver stopped immediately and cooperated with police, and no citations were issued. Under Arizona laws that generally favor drivers over walkers, pedestrians on shoulders of roads where no sidewalks are present are supposed to walk *facing* traffic, presumably so they can avoid approaching vehicles rather than vice versa.

An hour later, in Desert Hot Springs, California, Edward Manning, a forty-four-year-old father of four, was struck and killed by a passing car on the same route he walked every day to and from home—a residential avenue in this resort town notorious for poor lighting, lack of crosswalks, and drivers who routinely exceed the forty-mile-per-hour limit. A city official conceded in a press interview that lighting and speeding were regional problems that needed to be addressed, but opined that, with scant public money available for improvements, pedestrians needed to be more wary. A few hours after that fatal encounter, a fifty-four-year-old Alabama woman was struck and killed by a GMC Yukon sport-utility vehicle as she took an early-evening walk down a road in the border town of Donna, Texas.

The exact time Tony Ulloa was mowed down in New Braunfels, Texas, by a hit-and-run driver is unknown. The person who rammed into the sixty-five-year-old retired factory worker didn't stop and didn't leave behind much in the way of evidence except for a broken man lying in the road, groaning and bleeding. Ulloa had just taken up jogging to stay fit in his retirement.

He had nearly completed his usual course, which had taken him down the frontage road next to Interstate 35, just five minutes from his home. How much time passed after the hit-and-run driver fled, and what chance of survival he might have had with immediate care, are questions that plague the members of his large family—who say they are most haunted by the knowledge that he died alone when his whole family was so close by.

Overall, sixteen pedestrians are killed by motor vehicles every day in America. Another 466 are injured daily.[16] Only a third of the deaths and injuries are even arguably the pedestrians' fault: in 2013, 20 percent occurred from pedestrians running into the street, 7 percent involved improper crossing of the walkway, and 6 percent were from pedestrians standing, lying, playing, or working in the street. The majority are more like the collision that claimed Edward Manning's and Tony Uloa's lives, unintended but nonetheless the result of driver carelessness, indifference, or worse—as avoidable as they are commonplace.

A substantial minority of pedestrian traffic deaths are hit-and-run accidents—about one in five. Hit-and-run crashes of all types (car versus pedestrian, car versus bike, car versus car) have been on the rise for years and are ridiculously common in some cities where sprawl and readily accessible freeway systems make getaways easy. In recent years, a data search by the *LA Weekly* found, nearly half of all car collisions in the City of Los Angeles were hit-and-run cases, and the death and injury rate from hit-and-runs was four times the national average.[17] In 2014, 27 deaths and 144 serious injuries resulted from hit-and-run collisions with pedestrians, bikes, and other cars in Los Angeles; all reported cases of hit-and-runs (including parking lot scrapes and fender benders) in the city topped 20,000 that year. Arrests were made in less than 20 percent of the cases; more than half of reports were not investigated at all.[18]

The last year in which a comprehensive look at the problem nationwide was made, 2011, there were 1,449 fatal hit-and-runs in the U.S.[19]

Attitudes about what to do when the mix of driver and walker goes wrong have changed markedly over time. In the 1920s, as cars first became commonplace on city streets previously dominated by men and women on foot, rapidly escalating numbers of pedestrian injuries and deaths led to public outrage. There were demands for reform and pedestrian protections, occasional riots and acts of vigilantism against cars, and massive parades to protest car violence. Secretary of Commerce (and future president) Herbert Hoover launched a federal probe of the problem.

"Nation Roused Against Motor Killings," screamed a headline in the *New York Times* about Hoover's investigation on November 23, 1924. The story's illustration spoke volumes: a picture of a skeletal, caped figure of the Grim Reaper driving a giant roadster, crushing hundreds of children beneath its wheels. In explaining "the alarming increase in automobile fatalities," the opening of the news story warned *Times* readers that cars were a greater menace than World War I: "The horrors of war appear to be less appalling than the horrors of peace. The automobile looms up as a far more destructive piece of mechanism than the machine gun. The reckless motorist deals more death than the artilleryman. The man in the street seems less safe than the man in the trench."

The prevailing view at the time seemed to be that drivers who ran over people were by definition reckless simply by driving too fast or incautiously in areas where there were pedestrians about, particularly children. Carmakers came under pressure to install "governors" on engines that would prevent speeding over 25 miles per hour in highly populated areas. This turns out to have been a pretty good speed to single out. Subsequent research has shown that the vast majority of pedestrians survive crashes in which the vehicle is traveling under 20 miles an hour, while most victims die when the speed exceeds 40 miles an hour.[20]

In a barrage of marketing and lobbying, car manufacturers, auto dealers, auto clubs, and related industries in that era not only fought off such requirements as "antiprogress" and an affront to personal freedom, they succeeded in reframing the debate to shift blame away from drivers and onto pedestrians. This is when the term "jaywalkers" came into prominence with public and press, often accompanied by cartoon images of fools and bumpkins wandering cluelessly into traffic. Crashes began to be referred to less as innately reckless and more as blamelessly accidental—or due to the recklessness of pedestrians for failing to look both ways before daring to walk near moving vehicles. Next the idea that pedestrians were impediments to progress, travel, and commerce—and therefore should be confined to crosswalks and punished for crossing "against traffic"—came into vogue. Cityscapes were reengineered and traffic signals installed to manage rising car traffic, but the moves also had the effect of corralling pedestrians.[21]

This reframing of the relative rights of drivers and pedestrians began before the Great Depression and continues to dominate current law, street behavior, and thinking. Drivers today have little patience for pedestrians who "impede" them, and rules that allow right turns on red lights have forced pedestrians at crosswalks to hesitate before stepping from the curb because of the risk of being run down by heedless or distracted drivers, despite the fact that pedestrians have the right of way. Speeding is a principal factor in car-pedestrian crashes, yet a majority of drivers routinely exceed posted speed limits,[22] with many reporting that they find it impossible to keep up with traffic flow without speeding.[23] Engineers routinely create streets and roads that encourage this with designs appropriate for speeds far in excess of posted limits. When a child runs into the street and is struck by a car, the prevailing sentiment today, unlike sixty years ago, often places blame on the child's parents for negligent supervision. In Denver, there was widespread uncertainty over whether to view four-year-old Austin Strasser's death while legally crossing a street with his mother a crime or an unfortunate accident that could have happened to anyone. This confusion lingered even after the driver admitted to driving so carelessly it killed a boy. The antipathy is understandable: the driver who plowed into a mother pushing a stroller in a Denver crosswalk did something terrible, but did she do anything unusual? What driver hasn't turned the wheel and pushed the accelerator in a moment of inattention or impatience or bad decision-making? The explanation for little Austin's death, that the driver was blinded by glaring sunshine, is no excuse. The proper response when visibility is poor at an intersection where pedestrians are present is for the driver to stop until certain the way is clear, not to plunge blindly into a crosswalk. But other drivers make bad choices all the time in all sorts of circumstances. Nine hundred ninety-nine times out of a thousand, nothing bad happens. The driver—and the Strassers—were just the unlucky exceptions. Harsh jail sentences for doing something everybody else does daily seem hypocritical to many. But trivializing the carelessness that killed a child is no solution, either. So far this conflict between rules, practice, and decency has yielded a moral paralysis—and decades of a deadly status quo.

Some high-profile programs in major cities such as New York and Los Angeles seek to shake society free of this torpor, to make the world safer for pedestrians with lower speed limits in select areas where walkers and cyclists abound. The slogan "Vision Zero"[24] is often used to describe such efforts, which aspire to create a human environment in which there are zero traffic deaths. So far, the U.S. efforts on this score, and on traffic safety in general, seem to lag far behind the European programs they emulate—and have been plagued by pitched political battles and mixed messages as well. Drivers simply don't want to slow down, and all too often, they are enabled rather than discouraged, even in Vision Zero cities. The police departments in Los Angeles and New York responded to a rash of pedestrian deaths in recent years not by cracking down on bad drivers but by stepping up the issuing of very expensive tickets for jaywalking.

Another deadly pattern emerged on February 13, beginning with Heidi J. Springer. The fifty-two-year-old nurse-anesthetist at the Cleveland Clinic was ejected from her BMW X5 SUV after she drifted from her lane, overcompensated by jerking the wheel when she realized what had happened, then struck the concrete center median. Pedro Padron Sanchez of Hamilton, Alabama, lost control of his Chevy Silverado pickup in similar fashion when he wandered off Highway 253 in a moment of inattention. He, too, was hurled from his truck as it overturned. At almost the same time, an eighteen-year-old driver was thrown from his car after taking a sharp curve too fast and crashing through a farm fence, rolling the car four times. Christopher Short, meanwhile, drifted off a rural road at high speed and hit a ditch along Louisiana Highway 568, not far from his hometown of Waterproof. Short's Chevy pickup became airborne with such momentum, it crossed two lanes of traffic before landing, overturning several times on impact, and throwing him from the car. Short died on his nineteenth birthday.

Each of these crashes—and they were not the only ones to follow this pattern this day—had three things in common. They were fatal. The drivers were hurled from their vehicles with terrible force, inflicting catastrophic injury. And none of the dead were wearing seat belts.

The physics of car crashes are brutally simple: unrestrained people, pets, and objects turn into missiles inside cars during rapid deceleration. Isaac Newton first explained this phenomenon in 1687 with his First Law of Motion, and the merciless physics never change: if a car going a mere 30 miles per hour crashes to a sudden stop, everything that's not anchored in place continues to move forward at the same speed, striking anything around them with tremendous force. For the average adult male in the U.S., that force can have an effect roughly the same as dropping a twelve-ton weight on his head. That's more than twenty times the force of a professional boxer's best roundhouse punch. And so people fly through windshields. Their bodies bend or shatter steering wheels. Passengers in the backseat fly forward into the people in the front, injuring or killing them along with themselves. And if you're holding a baby in your arms, the infant will seem to weigh hundreds of pounds and be torn from your grasp, impossible to hold on to during impact.

And that's at 30 miles an hour. At 60 miles an hour, people fly through car windows and windshields like cannonballs. The official cause of death in such cases—and the single most common finding by coroners working fatal car crashes—is termed "blunt-force trauma." This is the coldly clinical term for the internal and external injuries a human body sustains when it is turned into a high-speed projectile striking metal, road, rock, tree, or ground. The reality is much messier than the term.

Seat belts are the single best way to avoid turning into a human missile during a car crash. The statistics on this are undeniable: the surviving passengers in fatal crashes studied in 2013 were wearing seat belts 84 percent of the time.[25] Only 16 percent of the survivors in those fatal crashes were not buckled up. There are always anecdotes of people who survived because their lack of a seat belt allowed them to be "thrown clear." That happens from time to time, but far more often the unbelted are thrown to death.

Seat belt use has steadily improved during the last half century. Overall, 87 percent of Americans say they wear seat belts when driving or riding in a car. This rate varies depending on the region and whether a particular state imposes strict fines for not wearing a seat belt. The West led seat belt use in 2014 with 95 percent of car occupants buckling up. The South was next, matching the national average of 87 percent, while the Northeast and Midwest trailed with 83 percent of car occupants using their seat belts.[26]

Men are 10 percent less likely to wear their seat belts than women, and rural residents are 10 percent less likely to buckle up than their urban counterparts.

Shortly before midnight, a married couple and their three-year-old daughter were all killed in Lonoke, Arkansas, when their Mercury sedan skidded and screeched off the road and into a tree after being rear-ended by a much heavier SUV. Glenna Michelle Wright, thirty-eight, who had been driving; her husband, Aundrey "Bucky" Wright Sr.; and their daughter Aunaysia Wright, all died at the scene, about forty miles away from their home in Stuttgart, Arkansas. The driver of the Mercury Mountaineer SUV was uninjured.

This mismatched weights and sizes of the two vehicles—one of several such mismatch crashes this day—illustrates a deadly trend that emerged in the early nineties in the U.S. After two decades of steadily dropping traffic fatalities on America's streets and roads, the numbers started climbing again. The shift coincided with another change: the rapid rise in popularity of a new type of passenger car—the sport-utility vehicle, a bigger, heavier car that was actually classified as a light truck (the same classification as a pickup truck).

The confusing part of this: heavier cars are supposed to be safer, not more dangerous, yet traffic deaths were going up.

Researchers soon figured out what was going on: vehicle owners were, according to University of California, San Diego, economist Michelle J. White, "running an 'arms race' on American roads by buying increasingly large vehicles."[27]

The SUVs that had become so popular at the time were, in fact, safer for drivers and passengers inside them, White reported in a 2004 paper. She found SUV occupants were 29 percent less likely to be seriously injured in a collision with a smaller car, irrespective of who was at fault in a collision. However, the reverse showed the high cost of that increased safety: the small car occupants were 42 percent more likely to be seriously injured in the same crash. Again, it didn't matter who was at fault. The occupants of the lighter vehicle were more likely to be toast.

And when all the various types of cars and traffic collisions were taken into consideration, White found that for every crash death avoided inside an SUV or light truck, there were 4.3 additional collisions that took the lives of car occupants, pedestrians, bicyclists, or motorcyclists. The supposedly safer SUVs were, in fact, "extremely deadly," White concluded.

She calculated that the safety benefit of replacing light trucks and SUVs with conventionally sized and weighted passenger cars would be "similar in magnitude to the benefit of seat belts."

The simple bottom line of this: heavier cars make most people more likely to die. If we all drove lighter cars, we'd all be much safer.

But the popularity of SUVs and the newer, similar class of vehicles known as "crossovers" has continued to rise, although some newer models do not run quite as huge as the original versions. In

2014, for the first time, SUVs and crossovers took a larger share of the American car market than sedans. Cars of all kinds have grown heavier as well in the last forty years: the original Honda Civic, a compact car, debuted in 1973 at 1,500 pounds but now weighs in at more than 2,800 pounds.

Expanding on White's work, a more detailed study out of Berkeley looking at the effect of vehicle weight on safety found that for every additional 1,000 pounds in a vehicle's weight, it raises the probability of a death in any other vehicle in a collision by 47 percent.[28] The added cost to society of overweight vehicles is $136 billion a year—costs that the owners of SUVs do not bear, the study's authors concluded. In order to make up for the added death and injury their vehicles cost the rest of the country, their fair share of the gasoline tax would have to be raised from the current 18.4 cents to $2.17 per gallon, the authors calculated.

Imposing this charge as the cost of using an inherently more deadly vehicle would, advocates argue, amount to removing a subsidy and allowing market forces to take over—which would likely dry up demand for heavy vehicles. But for a country that professes to believe in the power of markets more than public subsidies, there is absolutely no will or interest in having drivers pay the true cost of their choices. The U.S. government has not raised the gasoline tax since 1993.

On a more positive note, vehicle obesity is trending down for other reasons. After weights rose 26 percent overall for all vehicles between 1980 and 2006, government mandates to increase fuel economy are slowly nudging car weights in the other direction. The average U.S. passenger vehicle weighs in at just under two tons. Automakers are, if very cautiously, looking at substituting lighter aluminum and carbon-fiber composites for some heavier steel parts, which theoretically could cut vehicle weight in half with no loss of safety protection. Other trends, such as the arrival of fully autonomous cars, could shift Americans toward even smaller vehicles because virtually every crash that took place on Friday the thirteenth could, theoretically, have been avoided by replacing human drivers with robotic ones. The arms race toward big, heavy cars as the "safer" choice would end.

Brian Bayers, newly elected as Elk Creek magistrate for Spencer County, Kentucky, ran through his morning routine on February 13 like any other workday. He got his eighteen-month-old son, Jackson, dressed, fed, and ready for preschool. With his wife already off to work this frigid day, Bayers decided to dash outside and back his pickup truck into the driveway, where it could sit idling while the heater warmed the interior. When he returned to his house, he saw the front door he thought he had left closed now stood wide-open. Worried for his son, he ran inside, looking and calling for Jackson, but he could not find the boy. In a panic, Bayers ran back outside, fearful that, if his son had somehow gotten past the latched door—something he had never been able to do before—Jackson would be at risk of falling into a pond on the family's rural property. Then Bayers spotted the crumpled form of his only child beneath the pickup.

The toddler had somehow gotten outside right behind Bayers, trying to follow his dad out the door. He had trundled into the truck's blind spot as Bayers backed up. Jackson had been knocked over and run over by the front tires as the pickup backed into place. He died instantly.

Bayers called 911 for help, knowing there could be no help, then called his wife, Amanda, who rushed home from work so they could hold and rock their only child, whose pictures decorated seemingly every available surface in their home.

When backing up, motor vehicles don't have a "blind spot," as many drivers believe. They have a blind *zone*. The Kansas-based advocacy group Kids and Cars has made this message a key part of its crusade to prevent backover collisions. One illuminating poster the group distributes shows a photograph of a black SUV next to a house, with sixty-two preschoolers sitting cross-legged behind it, covering most of the driveway. None of the children are visible to the driver behind the wheel.[29]

The National Highway Traffic Safety Administration estimates that the inherently poor visibility behind most cars and trucks, coupled with driver complacency about the risks of backing up, leads to 210 deaths and 15,000 injuries a year. Nearly a third of the dead are under five, and another quarter are adults over seventy. Kids and Cars estimates that fifty children under the age of fifteen are backed into or over every month, with forty-eight of them requiring emergency care and two dying.

Backup video cameras for cars that eliminate the blind zones have been available for decades. (Prototypes were first demonstrated in the 1950s.) Safety advocates and parents of children who died in someone's blind zone have campaigned to make them as ubiquitous as safety belts since the turn of this century, and federal legislation in 2007 mandated backup cameras in all new cars,[30] though implementation has been repeatedly delayed as the deaths and injuries continue. They finally were scheduled to be installed on all new vehicles under 10,000 pounds sold in the U.S. by May 2018. The federal rule calling for this was only finalized in 2014.[31] Some automakers are already voluntarily providing backup cameras in their cars ahead of the requirement.

Cameras alone can reduce but not cure the backover risk, because drivers are still required to pay attention to the video display. Only an automated system that overrides the driver and prevents collisions by braking the car when an object is behind could do that. That technology exists—several new models of big-rig trucks on the road now have similar forward-looking collision avoidance systems in place already—but there is no mandate to put it on passenger cars.

Meanwhile, Brian Bayers agonizes over what might have been done to save his son from death, and other families from the same devastating loss and guilt. "What if I had a back-up camera on my vehicle? What if I had my window rolled down?" he said during a wrenching television interview he granted in order to warn other parents of the danger he never thought about before.[32] "I think: what if I just picked my child up and carried him with me to my car?"

Fridays are usually among the worst days for drunken driving. Friday the thirteenth was no exception, marked by a litany of people killed for doing nothing more than being on the same road with a drunk—for doing nothing more than trusting in the choices made by every other driver on the road.

At 2:00 p.m. outside Pittsburgh, one man was killed and two other people injured when a sport-utility vehicle made a left turn in front of an oncoming car on Pennsylvania Route 56. The crash turned into a police manhunt when the alleged driver of the SUV, thirty-four-year-old Jeremy Jonathan Blystone, ran from the scene. Blystone had a prior conviction for drunk driving and was wanted by police in three states.

The other driver, Thomas Pater, a sixty-one-year-old Vietnam veteran and local farmer, died in the crash. A passenger in each car was seriously injured. A three-hour police search using bloodhounds and a police helicopter ended in the nearby town of Apollo, where police caught up with Blystone after he stopped in a store, seemingly without a care in the world, to buy a pack of cigarettes.

A few hours later in Kenneth City, Florida, sixty-five-year-old Mark R. Ehrhardt died while crossing Fifty-Eighth Street a few blocks from home, run down by a Jeep Grand Cherokee SUV. The driver, forty-year-old Troy E. Donnelly, was arrested for driving while intoxicated and manslaughter after sheriff's deputies reported observing signs of impairment and Donnelly refused to submit to a breath test for alcohol. Donnelly had three previous convictions for drunken driving, the most recent in 2004.

And at 10:30 p.m., local entertainer Shane "Shaggy" Authement was making the two-mile walk from Marty J's Bayou Station, a truck stop and bar in Chauvin, Louisiana, to his family's home in the neighboring town of Montegut. Shaggy enjoyed partying as much as anyone and more than most, but the affable twenty-eight-year-old had one inflexible rule: no drinking and driving. He wouldn't do it himself. He wouldn't get in a car with someone else doing it. And he took away friends' car keys so they wouldn't do it. In keeping with his policy, Shaggy was walking home along the shoulder of Louisiana Highway 58 when a Toyota Camry driven by an alleged drunk driver struck and killed him. The fifty-three-year-old driver, from the same town as the conscientious entertainer, was charged with vehicular homicide.

Drunken driving, despite decades of tougher laws, police crackdowns, random checkpoints, and public awareness campaigns, remains the number one cause of traffic deaths. This runs counter to the widespread perception that public attitudes have turned sharply in recent years against driving drunk (after a long history of lax enforcement), so much so that the term "designated driver" has become a part of the everyday lexicon. Yet the statistics that document DWI carnage have stubbornly refused to budge in recent years. Ken Kolosh at the National Safety Council calls it "this brick wall we've hit" and identifies it as one of the main obstacles to reducing the car crash death toll going forward.

Consider these gruesome statistics:

- The average drunk driver has driven drunk *eighty times* before his or her first arrest.
- Every two minutes a person is injured in a drunk-driving crash.
- Costs directly associated with drunk driving in the U.S. are $200 billion a year.
- One-third of drivers arrested for drunk driving are repeat offenders.
- The age group most likely to drive drunk is twenty-one to twenty-five (nearly a fourth of all cases).
- On weekends, 31 percent of all fatal crashes involve alcohol. On weekdays, drunk driving is a factor in less than half that amount (15 percent of all fatal crashes).
- Men are twice as likely to drive drunk as women.
- More than 29 million people admitted to driving under the influence of alcohol in 2012—more than the population of Texas.
- The lifetime odds of being involved in a drunk-driving crash are two out of three.[33]

It is true that the number of drunk-driving fatalities has fallen dramatically since their high point in the seventies, when more than 60 percent of all traffic deaths involved alcohol. Since those days, when traffic deaths from all causes peaked at 54,589, several major changes in car design and the legal drinking age combined to bring that number down.

A spike in alcohol-related crashes occurred in the seventies after many states lowered the legal drinking age to eighteen. The trend reversed after the ages were raised again in the eighties and nineties, driven in part by the lobbying and public awareness campaigns by the Mothers Against Drunk Driving organization.

Separately, a 1968 federal law[34] mandating seat belts in all vehicles except buses and, later, state laws requiring people to actually wear them or face stiff fines, reduced traffic deaths (though not traffic crashes) dramatically. So did subsequent collision safeguards: air bags and more crashworthy car frames and bodies. Other technologies actually prevented crashes by compensating for human error, such as antilock braking systems that became commonplace in the late eighties, lowering the risk of fatal multivehicle collisions by 18 percent and run-off-the-road fatalities (particularly loss of control on curves) by 35 percent. All of which meant that crashes, drunken or otherwise, were being survived that in the past might have been fatal.

Combined with raising the legal drinking age, these changes had cut the number of drunk driving deaths in half by the early 2000s. The drunk-driving fatality rate has been hovering in the range of one-third of all deaths ever since—which means about 12,000 deaths caused by drunk driving in 2014.

There are devices that would add a few hundred dollars to the price of a new car that prevent them from being started by a legally drunk person based on breath analysis. This technology has been used successfully in some states where judges are empowered to make it a condition of a convicted drunk driver's sentence, but this is done in a minority of cases, even those involving repeat offenders. The technology has become so simple that comparable pocket-size devices are on the market that can be clipped to a smartphone, so technology is not the barrier. Touch sensors that can measure alcohol in the blood by "sniffing" the skin are also being tested and show promise. But proposals to use drunk-driver lockout devices more widely—or to have them installed in all new cars—have garnered little support and gone nowhere, although the payoff is potentially huge. Putting them in all new cars would be something of an inconvenience, but over fifteen years, as they gradually became ubiquitous, the gadgets would prevent an estimated 59,000 deaths, 1.25 million crash injuries, and $349 billion in crash injury costs—much, much more than the cost of adding the devices.[35]

There is one encouraging change in progress. Some evidence suggests that ridesharing services have reduced drunken driving arrests, at least among men and women under the age of thirty, the most frequent ridesharing customers. Every rideshare driver knows the busiest time and place for picking up a customer: outside bars at closing time. Data collected by rideshare industry leader Uber for the seventeen markets it serves in California showed a 6.5 percent decline in drunk-driving crashes involving drivers thirty years old and younger after Uber entered the market.[36]

The final fatal crash of February 13, 2015, began with a dropped mattress in the fast lane of the 55 Freeway in Santa Ana, California. At 11:40 p.m., the wayward mattress triggered a four-car chain reaction of collisions that in turn led to a hit-and-run, a manhunt, a drunk-driving arrest, two people injured, the death of one driver, and a mystery as to where the mattress came from in the first place. The northbound side of the busy 55, also known as the Costa Mesa Freeway, stayed closed until sunrise the next morning as the California Highway Patrol photographed skid marks, measured the trajectories of the cars, and slowly tried to piece together what happened beneath the harsh glow of emergency floodlights.

Cynthia Brock, a fifty-three-year-old resident of Costa Mesa, was the first driver to encounter the mattress, dropped by an unknown driver in one of the freeway's left lanes. The mattress caught beneath the front end of Brock's 1982 Toyota Celica, which skidded out of control and slammed into the concrete center divider, then slid to a stop sideways in the fast lane. A Ford van immediately broadsided Brock's helpless car, striking the driver's side full on. The van driver, whom the CHP identified as nineteen-year-old Sherwin Ali Sabzerou of Irvine, California, managed to pull his van to the shoulder five hundred feet up the road, then allegedly fled the scene on foot. Meanwhile, the van's impact pushed Brock's Celica into the path of a 1999 Toyota sedan, which rear-ended her, spinning the car around so that it was facing the wrong way. Then it was struck head-on by a 1995 Honda.

All the vehicles were badly damaged, but Brock's looked like it had been run through a junkyard compactor after four collisions with cars and one with concrete. The driver's door hung by a thread. The Toyota's roof had been crushed and the front end accordioned. The wheels hung askew, and broken glass had been sprayed everywhere. Brock died before the CHP arrived. Two occupants of the other cars were hospitalized as well, with both expected to recover.

Police found Sabzerou at his home and arrested him on suspicion of drunk driving. He was later charged by prosecutors with hit-and-run causing injury or death and vehicular homicide. However, the ultimate culprit who caused it all, the person who dropped the mattress, has not been identified.

Objects dumped on freeways are a daily occurrence throughout California and the nation. The California Department of Transportation maintains storage yards filled with retrieved roadway debris and maintains special crews to work with the CHP to pick up objects on the freeways, a chore that is among the most high-risk jobs in town, because there is never a time of day or night when traffic is not present. Toilets, televisions, flat tires, chairs, suitcases, refrigerators, and all sorts of bedding, including mattresses, turn up regularly on freeways, where speeding vehicles attempt to dodge them until they can be cleared. There's also a chronic problem with debris left over from scavenging gangs, who steal miles of copper wire from freeway lighting, signals, signage, and even the metal embedded in the roads for traffic sensors and water pumps, leaving traffic hazards in their wake, costing the state tens of millions of dollars to replace.[37] It's a constant game of chicken, of taking away vital parts of the freeway system and dumping unwanted trash in its place, turning engineering marvels into minefields where even a mattress can have the effect of a bomb.

As with backup cameras, forward-looking collision avoidance systems already exist and could have prevented this entire fatal pinball machine effect. Airliners have had them for decades—the result of painstaking crash investigations that proved they were needed and would be worth the cost. But that's air travel. The mattress would not be so painstakingly investigated. The careless, clueless killer who left it on the freeway would bear no consequences. The mandate, the public pressure, the will to provide cars with the same protection that airliners have, doesn't exist. Even the oceangoing shipping containers filled with sneakers or lawn mowers or scrap metal enjoy more protection from damage than the containers that carry people—that carry us and those we love. Cars are different.

And so ended one day of traffic carnage. Every day begins and ends the same as February 13, with ruin and road closures, the consequences of one day's wreck carrying over into the next morning. (A list of the day's crashes can be found in the Appendix.)

The sheer volume of one day's worth of car death and injury forms a kind of protection for even the most reckless drivers. There are just too many of them. Most drivers speed. Phone use behind the wheel is endemic. Only a tiny fraction of drunk drivers get caught. Imposing true and reliable accountability for putting others at risk on streets and highways would be like policing a tidal wave. There simply is no money, manpower, or public demand for fully investigating car deaths with the same vigor applied to aviation or food safety or epidemics.

Two deaths from Ebola in 2014 sparked a national panic. One German airliner crashed purposely by its copilot in 2015—a truly unique event—generated worldwide discussion of how to build failsafe systems that would prevent that most rare of events from ever happening again. It is a curious truth of human behavior that the rare risks—the ones least likely to harm us—are the ones we most fear and try hardest to conquer. This reaction is instinctive rather than reflective. The far greater risks we face daily, such as driving fast and distracted on crowded freeways, are perceived as normal and routine, becoming functionally imperceptible to the brain unless we consciously dwell on them. Just as aberration pushes us to overestimate risk, habit pushes us to underestimate it wildly. One car death every fifteen minutes, and one crash every three seconds, become little more than white noise to us. We couldn't get in our cars every day otherwise. And we certainly couldn't cruise down 65-mile-per-hour freeways with our babies strapped into car seats that offer little protection above 35 miles per hour. Habit conquers all.

There are also practical barriers when it comes to holding bad drivers accountable for their risky and poor choices. Truly enforcing the rules of the road with our current infrastructure and cars poses far too great a burden to police, beyond attempts to round up and punish drunk drivers with occasional and controversial checkpoints that rarely lead to arrests (although they may have a deterrent effect).

When the authorities try to crack down on sober motor violence, the efforts are even more controversial and are often met with community uncertainty, even outrage. When police arrested a New York bus driver for running down a schoolgirl in a crosswalk on this very morning of February 13 (she was seriously but not fatally injured, one leg badly mangled when she was pinned by the bus), the *New York Daily News* decried what it saw as mistreatment of one of the city's bus drivers. The head of the transit union protested this enforcement of the city's new Right of Way Law as "outrageous, illogical and anti-worker" while branding the head of a city street safety advocacy group "a progressive intellectual jackass."[38] The same union previously launched a work slowdown when another bus driver faced sanctions for killing a seventy-eight-year-old woman in a crosswalk in December 2014. All this occurred because police sought to enforce *misdemeanor* charges in cases of pedestrians who were run over in crosswalks where they had the clear right of way. The offenders were not just run-of-the-mill drivers hurrying to work and succumbing to regrettable, though perhaps understandable, human failings. These were professional bus drivers with a $67,444 annual base salary plus generous overtime,[39] whose job includes, above all other considerations, a duty for safe, alert, and lawful driving, in a city with more pedestrians crossing streets than any other municipality in America. Yet even these halting attempts to hold drivers accountable for the street injury and death they inflict on innocents are met with resistance and doubt.

In Springfield, Oregon, later in February 2015, a sixty-eight-year-old man was accused of running a red light on State Route 126 (which serves as Main Street in Springfield), killing three children and critically injuring their mother as they crossed. After months of soul searching, investigation, and dithering, the authorities found the driver was neither drunk, nor speeding, nor on his cell phone when he decimated a family, and so no charges of any kind would be filed.[40] It was just an accident, the local newspaper editorialized, in support of the official findings, an accident on a street with homes and families and a forty-mile-per-hour speed limit, in a world where most vehicle-pedestrian collisions at that speed bring catastrophic results. The two cardinal rules of driving—stop for red lights and stop for kids—had been broken by a driver who could not or would not behave as if he was in charge of a machine as deadly as a gun, and that every decision or incident of thoughtlessness behind the wheel really was a matter of life and death. We have trivialized the dangers and risks of driving, just as we have trivialized the inevitable, fatal results. The state newspaper editorialized, after much hand-wringing, that "'accident' is the only way to accurately describe what unfolded at that intersection . . . Just a tragic accident."[41]

In a way, such antipathy makes perfect sense, because a terrible truth lies at the heart of our star, the car: the American system of roads and wheels is performing exactly as it was designed to perform.

Charles Marohn, who identifies himself as a "recovering traffic engineer" and who founded the Minnesota-based non-profit group Strong Towns, publicly rails against his profession for knowingly designing traffic systems where deadly interactions between cars, pedestrians, and intersections are inevitable.[42] Who, Marohn asks, is most responsible for that fatal crash in Oregon? Is it the driver who was momentarily inattentive and mistook a red light for green? Or should blame be laid before the engineers who make residential streets with design speeds 20 or 30 miles per hour faster than the posted speed limit? Or the policy makers who allow posted speed limits high enough in residential areas to be catastrophic every time a pedestrian collision occurs? The street where the three children died carries four lanes of traffic with a center turn lane—a design that invites risky speeds in a family neighborhood.

"Speed is seductive," Marohn observes. "We engineer for high performance. Can we then blame drivers for taking advantage of that engineering? We know they will. . . . This is indicative of our incoherent approach to streets and roads."

Marohn argues that the distinction between streets and roads has become blurred over time, a confusion at the heart of motor violence. Streets are supposed to be platforms for creating wealth, he argues, while roads exist to get people from Point A to Point B as quickly as possible.

Great streets thrive on complexity, which today can encompass the safe (and slow) mix of pedestrians, cyclists, cars, buses, trolleys, delivery vans, schoolchildren, shoppers, business people—the classic Main Street mix. Think of any street where you love to walk or window-shop or sightsee, where you slow or stop your car to turn and park and no one angers, honks, or glares, and you get the idea. Great streets build business, society, prosperity, and strong towns.

Roads serve a completely different purpose. They do not tolerate complexity. Think of freeways. They have barriers that prevent turning. They have no traffic signals or crosswalks. Pedestrians and bicycles are banned. There are no roadside attractions to park and visit without getting off the freeway first. Roads serve the simple purpose of taking us and our stuff somewhere far and fast. They can take you to streets, but they can't *be* streets.

"We need both," Marohn says. "What we don't need is something that tries to be both."

Yet the modern landscape is filled with thoroughfares trying to be both. Marohn coined the useful term "stroad" to describe these often unsightly hybrids that arose during the age of postwar suburban sprawl and the car-centric traffic engineering philosophy that accompanied it. The fatal crash in Oregon took place on such a stroad—a conveyance that has the worst attributes of street and road while performing neither function well. Stroads offer commerce without walkability—fast-food drive-throughs, strip malls, big-box stores—but they usually lack the larger economic payoffs that great streets generate. At the same time, the turning, parking, and crosswalks bolted onto these fast, wide stroads slow down the Point-A-to-B traffic flow, imposing the cost of delay on drivers, encouraging them to speed even more when they are moving.

This uncomfortable mix makes stroads the scene of many crashes, particularly those involving cars hitting pedestrians. The official response—when there is any at all—is usually to armor up the roads with guardrails or barriers or, as in the Oregon case, to propose stepped-up police patrols. Rarely is the obvious and only effective solution imposed, Marohn says: making sure cars can't go faster than a statistically safe 20 miles per hour where significant numbers of pedestrians are present. That means having pure streets and pure roads, with hybrids taken out of the mix.

The problem is that stroads have become an enormous part of the American landscape. Converting them one way or the other would be a long and arduous project—and a controversial one. Pushing this viewpoint has turned Marohn from insider to political pariah in his own Minnesota community. His current cause there is to generate support to turn a disused highway that cuts through the center of his town of Brainerd into a bike and pedestrian-friendly great street. But too many people associate getting somewhere fast with prosperity, he says, ignoring the evidence that true streets—slow and safe streets—play an important role in creating wealth, and do so better than stroads.

"Until we can change that attitude," he says, "it is inevitable that we continue to have tragedy."

Today's combination of powerful cars, street and traffic engineering, and human behavior is designed to produce 36,000 deaths and 2.5 million injuries every year. That toll is not some unintended by-product of our personal transportation choices but the predictable and expected result of the design choices—and design defects—built into our transportation system and our human selves. Cars are convenient, fast, and mostly far bigger and more powerful than they really need to be, given that they drive around with only one person and mostly empty seats and cargo space 75 percent of the time. A death every fifteen minutes is part of the price tag for that convenience, size, and speed.

And here a strange dichotomy arises between man and machine that warrants attention.

Regulatory agencies, the legal system, advocacy groups, and the media leap into action when there is a design flaw in the *machines* we drive. On February 13, 2015, a day of normal human error causing normal traffic carnage, General Motors announced the recall of 81,000 passenger cars with a flaw in the power steering—a flaw that caused one crash and no death or injury. The previous year, General Motors recalled 2.2 million cars to repair faulty ignition systems that unexpectedly shut off the cars, leading to 124 deaths.[43] And in May 2015 the largest recall in history would be announced, involving 34 million cars in the U.S. from multiple carmakers (57 million worldwide), all of them equipped with flawed air bags made by the Takata Corporation of Japan. Fixing them will cost the company at least $2 billion for repairs and potentially billions more in litigation costs. The defective bags spray metal fragments during an accident, much like shrapnel from a grenade. Six deaths and more than one hundred injuries have been linked to this problem.

Recalls often require months, even years, of investigative and regulatory pressure in order to hold carmakers accountable. This is a considerable effort that, however vital, involves numbers of deaths and injuries so small they barely register statistically.

When it comes to the overall death and injury rate from driving, fault rests overwhelmingly with the choices of drivers, not the mechanical defects in cars and components.

But when the defect is in the human behind the wheel rather than the machine itself, the story takes a very different turn. There has never been a recall aimed at fixing cars so drunks cannot start them, or so drivers cannot exceed the speed limit, or to prevent cell phones from being used while cars are in motion. Cars currently are designed to allow, even enable, all three of these deadly behaviors, although affordable technology exists that could make all three impossible. Their omission can be viewed as design defects, too—defects in the human-machine-road interface, which kills far more than any bad air bags and faulty ignition switches. Undertaking a "recall" to correct any one of these human design defects could prevent 30 deaths and 2,200 serious injuries *every day*—more death and injury prevention than all the car recalls in history. Now that would be a true Vision Zero.

Instead, Americans have thrown up their hands and written off daily death and injury on a massive scale as the inevitable cost of mobility in the modern world. The predictable results of epidemic-scale fatal driving behavior and design are not viewed as defects but as "accidents," a choice of language that suggests being stuck down by someone carelessly misdirecting two tons of speeding metal is somehow equivalent to being struck down by a bolt of lightning. But they are not the same. One is largely preventable, while the other, not so much.

There is another difference: the odds of lightning killing a U.S. citizen in the course of his or her lifetime is 1 out of 136,011. The odds of that same person dying in a car crash: 1 out of 112.

Chapter 6

PIZZA, PORTS, AND VALENTINES

The ports are a mess—and I can't get work," the longshoreman turned Lyft driver complains. He has picked me up outside the conference on green ports of the future, but my driver is more concerned with the dysfunctional ports of the present. He pats the wheel of his black sedan. "So this is paying the bills for now, Lyft and Uber. A lot of the guys are doing it. Course, it's nowhere near the money we make moving cans."

Moving cans. That's what the longshoremen call their work loading and unloading shipping containers from cargo vessels. They move those cans back and forth to vast dockside yards—the terminals—to await loading on trucks or trains. Then other crews move fresh cargo back onto the ships. Stacks of empty containers accumulate in a kind of no-man's-land at the port, with many of them shipping back to Asia empty. Ships arrive filled with imports, but far fewer goods travel back in the other direction.

The container ships usually make a quick turnaround, because time literally is money in this business, a billion dollars of it in goods moving every day through the conjoined port complex of Los Angeles and Long Beach—the quickest and best shipping option in the U.S., according to its promotors. But in February 2015 the wait can be weeks instead of two or three days, so bad is the congestion and overload at the ports—delays in delivery that are grinding the door-to-door machinery into low gear nationwide.

Today twenty-seven cargo ships are languishing in backup anchorages, a flotilla of vessels bigger than half the navies in the world. Some of them have been waiting as long as ten days to get in. The congestion has multiple causes: too few trucks, bottlenecked roads, bridges in need of modernization, and, most of all, bigger ships with bigger loads arriving at the same old docks. But the most acute and controversial cause of congestion at the moment, and the one drawing the most attention and public ire on February 13, is a nine-month labor dispute.

A major sticking point in settling a new contract between the longshoremen's union and the shipping companies is the obscure but pivotal system of arbitration, which affects all aspects of work on the docks, from safety rules to wage disagreements to the protocol for taking restroom breaks. That's all it takes to cripple a hub of global commerce, and Tom,[1] a second-generation longshoreman from a family of dockworkers, is among the thousands out of work as a result. It's not a strike—not even an official job action, he explains as we drive toward home—but rather a policy of following every work rule to the letter, as if every driver on the road suddenly stuck to the speed limit. The shipping lines call it a work stoppage and they're furious, but the union claims rules are rules. The union, the shipping lines, and the terminal operators are all contributing to the impasse: there is blame aplenty for every side of this struggle that has hobbled all twenty-seven West Coast commercial ports. Together those ports handle almost 13 percent of America's gross domestic product and create 9 million jobs. But it's Tom and his fellow workers with little seniority, the women and men in their twenties, who are out of those high-paying dock jobs while the titans battle. So he drives for the ride-share companies, picking up fares instead of containers for about a fourth of his normal wages or less at the port.

"We want this to settle, but we're also not going to back down," Tom says. "This situation is not good for anyone."

This is an understatement. When ports that handle so much of the national economy slow down, their far-flung connections and importance in our daily lives become poignantly clear as the suffering spreads far. The impact radiates out from the port, affecting everything from farming to gas prices to car sales to the celebration of Valentine's Day. ("Don't even ask," Tom says, eyes rolling in half jest. "There's a chocolate heart shortage. It's a disaster.") Little is spared by the overload, not even our national staple, pizza. Just ask Domino's.

While my wife and I have an anniversary dinner at our favorite restaurant on Friday the thirteenth, our teenaged son stays home and orders up a pizza. He uses Domino's online pizza tracking app, which lets him design his own pie, informs him when the pizza is popped in the oven, and tells him who at our local Domino's franchise is doing the cooking and when his dinner is on the way to the house.

All of which disguises the fact that Domino's real business isn't pizza: it's logistics and transportation. The individually owned franchises—the actual restaurants, 11,600 of them worldwide—are in the pizza business. Domino's the corporation only owns about 300 of them, mostly to test new products and decor. The main profit engine for the Domino's company, however, lies in making, obtaining, and transporting supplies to its franchises. The dough, so to speak, is in the dough (and the sauce and the cheese and the pepperoni), about 63 percent of the $2 billion in revenues.[2]

The heart of the business in the continental U.S. lies in the sixteen Domino's supply chain centers spread around the country. For Southern California and the pizza my son scarfed up this day, the supply center is in an industrial park in the city of Ontario, where an affable pizza lover by the name of Don Fontana, vice president of the company's west region supply chain services, bemoans the shortage of pineapple slices caused by supply problems at the port.

"We've had to scramble to find alternative sources," he grumbles. The Hawaiian-style ham and pineapple pizza is one of the more popular combos in the region, Fontana says. (Regions vary: green olives are popular in New England, and Texans like beef on their pizzas.)

Pineapple is one of the few internationally sourced ingredients Domino's uses—Thailand and the Philippines are main sources—and so the port congestion has an impact here. It also affects other Domino's suppliers: the vast factory farms in California's Great Central Valley that grow tomatoes specifically for sauce, and the industrial sauce makers that set up next to them, peg their operations on the ports, too, because they are the world's leading exporters of the stuff. Their operations—and revenues—slow down because they can't move product out fast enough.

Fontana enjoys showing off this operation, the third busiest in the Domino's empire, serving 347 franchises. While in the warehouse area he wears a hard hat, which looks big on him. He's pretty thin for a guy who says he eats pizza several times a week, although he is a purist most of the time, preferring cheese and crushed red peppers as his only toppings. He waves enthusiastically at the twin two-story-tall tanks of flour, each containing enough to make a quarter million pizzas. They stand well away from the food preparation area, equipped with "fluidizers" that use air pressure to make the powdery flour flow through pipes and conduits like a liquid. They're filled four times a week by big-rig tanker trucks that pull up alongside the warehouse and pump flour through intake pipes on the outside of the building. No sacks of flour in this operation; it would take far too long and far too many sacks.

Everything at the Ontario plant revolves around goods movement, Fontana says, even the "supply chain center" designation, which sounds a bit cold and clinical for his taste. Each center used to be called "the commissary," a term derived from the company founder's Marine Corps background. "Everybody still calls it that," Fontana confides, although the official name changed about twenty years ago. "When we call and use the term 'supply chain center,' no one knows what we're talking about until we mention the word 'commissary.' "

The journey of my son's pizza starts at 4:00 a.m. in Ontario when the first of fourteen big rigs arrives with the day's supplies, starting with two truckloads of mozzarella. That's 2,736 fifteen-pound bags from cheese giant Leprino Foods' branch in Lemoore, California, 233 miles north and made from milk sourced from California dairies. Another truck arrives with 936 cases of sauce from TomaTek in Firebaugh, California, in the heart of tomato-growing country 278 miles north of Ontario. The Tyson Foods delivery brings pepperoni, sausage, ham, and salami in from the Dallas area, 1,400 miles, and chicken toppings out of Arkansas, 1,600 miles. Presliced onions and bell peppers come in from Boskovich Farms, just 100 miles away in Oxnard, California, while one of the top five toppings, mushrooms, arrives from Monterey Mushrooms in Watsonville, California, 361 miles distant. Flour originates in the wheat belt 1,500 miles away, but the mill that delivers it by tank trunk daily, Ardent Mills, a joint venture of food giants Cargill and ConAgra, is just twenty miles away in Colton. Salt is shipped in from Cargill in Wayzata, Minnesota, 1,900 miles distant, while sugar arrives from Cargill's Brawley, California, plant, just 163 miles away. Assorted deliveries of less frequently used toppings—garlic, anchovies, banana peppers, beef strips, and jalapeños—round out the offerings, along with frozen bags of preprepared pasta dishes and chicken strips and beef for sandwiches. Multiple deliveries of empty pizza boxes arrive throughout the day from Santa Fe Springs, 33 miles away, although they're made by a Georgia company 2,200 miles away, making them the most distant piece of the pizza puzzle other than pineapple. The various ingredients are parceled out to sections of the warehouse that are refrigerated, frozen, or kept at room temperature, where they await loading on outgoing trucks later that day.

The only freshly prepared pizza component is the dough, along with the fresh-baked sandwich rolls, whose enticing aroma permeates the commissary's baking area and occasionally wafts onto the warehouse floor. Everything in the Domino's supply chain center revolves around the dough-making operation cycle, which begins when the ovens start preheating at 5:00 a.m.

Domino's pizza dough has six primary ingredients that go into one of three giant mixing bowls at the plant. Each mixing bowl holds more than six hundred pounds of dough, consisting of flour, yeast, salt, sugar, water, and oil. A secret "goody bag" with a small quantity of Domino's proprietary flavors and dough conditioners is dumped in the mix, too, and giant stainless steel beaters go to work kneading the mixture under the supervision of dough master Julio Rico. When the mixing is done, the giant bowl is loaded on a clanking stainless steel lift that raises the dough about eight feet in the air and then overturns it into a cutting machine that extrudes dough cylinders like Play-

Doh, dumping them on a conveyor belt. The belt whisks the pasty-looking cylinders to a rolling machine that turns them into balls of dough ranging from baseball to softball size, depending on whether they are for small, medium, large, or extra-large pizzas. The dough balls shoot through a metal detector to make sure no twist-ties or bits of machinery contaminated the dough, then three line workers inspect, flatten, and pack the dough balls into one of the thousands of blue plastic trays that fill the facility in tall stacks.

By 1:30 p.m., the production phase of the day ends with the last dough run, whole wheat pizza dough for school lunches. The daily output: enough dough for 100,000 pizzas, plus 7,200 sandwich rolls partially baked, to be finished just prior to serving at the franchises.

At 2:00 p.m. the loading of the outgoing trucks begins. These are Domino-branded refrigerated big-rig trucks owned and maintained by Ryder and made by Volvo, with forward-looking collision avoidance systems designed to eliminate rear-enders. It's a first small step of bringing automation into the trucking side of the goods movement system. The trucks have been plugged in to the supply center's electrical system and cooling down to 36 degrees all day. Even the loading dock is refrigerated to protect the raw dough. The bulk of each trailer's interior space is taken up by the towers of stacked blue trays with their 100,000 dough balls, layered and mapped into sections based on the size of the pizza (medium and large are by far the most popular). The other ingredients—cheese, sauce, toppings, golden cornmeal to dust the pizza pans, napkins, and red peppers, in addition to cardboard pizza boxes—have to be crammed in around the all-important dough trays.

At 8:00 p.m., the first of the trucks departs for deliveries to the franchises, continuing in waves through midnight. Each truck has its own geographic area that might have twelve to fifteen stops, ranging from close-in deliveries in the LA metropolitan area, to franchises as far as the Arizona border, the Mexican border, and the ski resort at Mammoth Mountain, the most distant stop at three hundred miles and the only overnight run. The goal is to deliver the goods while the pizzerias are closed. The drivers have keys and put everything away, so the store is stocked and ready to start cooking the moment it opens for business.

Over the years, Domino's has had to build ever more amounts of time into its logistics system as traffic and overloaded infrastructure slowed down the arduous process of assembling ingredients and bringing our pizzas door to door. Before my son had even finished his pizza, the first truck was departing Ontario, headed to the coast and our local Domino's with the supplies for the next day's orders.

Friday the thirteenth proved to be a busy day for the transportation system of systems beyond the worlds of pizza and ports, offering glimpses of future disruption and present overload in a series of interconnected events.

In Mountain View, California, the dreamers behind the Google self-driving car project logged yet another day in their quest to navigate a million miles with nothing but algorithms behind the wheel. The search giant's modified robotic Lexus SUVs mixed it up with regular cars, bikers, and walkers on the busy streets around the Google campus without error—other than the fact that the robots were the slowest cars on the road. It seems there was a software glitch human drivers lack: Google cars refuse to treat the posted speed limit as a suggestion, and have been rear-ended as a result. "The least reliable part of a car," Chris Urmson, head of the Google car project, likes to say, "is the driver."

In Sacramento, meanwhile, the state's department of motor vehicles struggled to craft the nation's first rules of the road for a point in the future when driverless cars leave the current test phase and are fully loosed on the world for any and all who want them. The DMV's stated primary goal is to keep humans safe from the robots, although all evidence to date suggests that it might be better to make that the other way around. The regulators were stymied by the question of how to test the safety of the technology without stopping its development in its tracks. A key question was whether to follow the longstanding practice of letting the industry do the safety testing and report the findings, or to have independent safety testing and analysis by outside experts. Industry trust has been severely damaged in the wake of revelations that General Motors failed to disclose fatal ignition system defects for a decade,[3] and by the discovery in late 2015 that Volkswagen had installed illegal software cheats on its popular diesel cars to make them appear to be far less polluting than they really are.[4]

In Seattle, another robot story unfolded as the leaders of retail giant Amazon.com suffered a setback in pursuing their vision of mini-drones replacing human delivery trucks and drivers. The Amazonians were upset to learn that new federal flight rules would require commercial drone operators to remain within direct sight of their little flying minions. This precaution fits just fine with drone real estate inspections, crop dusting, and aerial filming on movie sets, and it's what current rules require of hobbyists who fly remote control planes. But it makes little sense for delivering packages, as it would turn today's one-person, one-truck operation into tomorrow's one-person, one-truck plus drone operation. Instead of saving money by cutting humans from the field, it would add to costs. If Amazon wants to pilot their drones all long-distance, Air Force bunker–style, it seems the company will need a swarm of lobbyists before it can unleash the drones.[5]

In Houston, commuters were wishing they could launch their own drone attack on state highway engineers as lines of cars crawled down a stretch of Interstate 10 that had been expanded from eight lanes to a colossal twenty-six—the $2.8 billion Katy Freeway project that was supposed to solve traffic forever. Six years after the expansion, a thirty-mile rush hour drive from downtown that once took under forty-seven minutes now takes more than seventy minutes on the widest highway in the nation if not the world. This is the build-it-and-they-will-come effect writ large, known to traffic engineers as the phenomenon of "latent demand or induced demand." Before construction began, light rail lines were suggested in place of some of the added car lanes in order to tame future congestion with a mass transit option, but the proposal was rejected as a waste of time and money. In California, transportation officials were chortling. Texas has made Carmageddon look good by comparison.

Residents of San Bernardino, California, however, woke to the sound of the suburban dream breaking. Their city began holding a series of town hall meetings to discuss possible outcomes of their city's bankruptcy—none of them pretty. Located in the desert east of LA, San Bernardino exists in its modern form as the epitome of car-dependent suburban sprawl, its entire economy built on cheap land, subsidized highways, and subsidized free parking, and a landscape of immense warehouses that take advantage of the cheap land. These are the policies that created America's car-centric landscapes since the 1950s. When San Bernardino's turn came at the start of the millennium, new residents flocked to vast tracts of new homes with big yards and pools, priced half or less than similar houses in LA. The price also included dependence on hour-long (or longer) commutes, but as long as real estate prices and demand kept spiraling up, the trade-off seemed to make sense. The inability to sustain this version of the American dream became apparent when the housing bubble burst, jobs dried up with the recession, and whole San Bernardino subdivisions went belly-up as the foreclosure rate hit 3.5 times the national average. Now the bankrupt city's financial ruin is second only to Detroit's. It has been said that Henry Ford's true invention was not the modern car but the modern traffic jam, although in San Bernardino it seems what he really invented was modern sprawl.

Elsewhere on February 13, the traditional annual announcement of sales for the most popular car models was made today, revealing just how much Americans still love their traditional internal combustion cars—and how fickle their embrace of more efficient alternatives has been. Nationally, the two top sellers were pickup trucks. In California, America's biggest vehicle market, where greener cars have traditionally done well, the Honda Accord conventionally powered sedan outsold the fuel-saving hybrid, the Toyota Prius, for the first time in years. This shift reflected a national trend away from hybrids and alternative-fuel vehicles, which Americans were dumping due to the 2014 plunge in gas prices. Sales of the most popular of hybrids, the Prius, declined 12 percent. Meanwhile, SUVs were making a comeback, too, although slimmer ones than in years past were leading the resurgence.[6]

A subplot in the story of temporarily declining gas prices unfolded today as well, in the form of two immense freight trains barreling cross-country toward depots in the Virginias. Given priority over trains bearing crops, cows, and passengers, each of these trains carried millions of gallons of crude oil from the boomtowns of North Dakota and points north. This oil came from the Bakken shale deposits recovered through the extreme form of drilling known as hydraulic fracturing, or fracking.[7] The rapid rise of thousands of such oil trains as a major new piece of the American energy delivery system accounts for America's unexpected rise to the world's leading petroleum producer in 2014, and it's those rail shipments that glutted the market and helped drive gasoline prices down. About 75 percent of this new oil from the Bakken moves by train.[8] But these two trains rumbling toward the East Coast, like an alarming number before them, would never reach their destinations. One would crash and burn on Saturday the fourteenth in an isolated area near the Canadian border; the other would make it as far as Mount Carbon, West Virginia, before nineteen fuel-laden cars would tumble from the tracks, burst into flames, and send fireballs rising into snow-filled skies. Hundreds of families had to be evacuated. Communities lost water and power. A new federal rule mandating safer tanker cars, long delayed and disputed by the rail industry, was still in the works at the time of the crashes. It will take years more to implement amid debate over the wisdom of running so much volatile fuel through cities and towns.[9]

The wisdom of operating massive oil refineries in the midst of neighborhoods, schools, and shopping centers would also become a source of scrutiny and concern because of a seemingly innocuous task under way today: the shutdown and maintenance of a massive oil-cracking tower. This normally routine task had set in motion a chain of events that would end in an explosion five days later. The blast, caused by an overpressurized antipollution system known as the fluid catalytic converter, took place at the 750-acre ExxonMobil Torrance Refinery, capable of producing up to 10 percent of California's gasoline supply. This steaming, flaring industrial outpost is surrounded by the affluent coastal communities of Los Angeles County collectively referred to as the South Bay, and the explosion rained down ash and debris on its neighborhoods. Four workers were injured, and nearby schools were forced to keep students huddled indoors. This was the fifth explosion at the facility since November 1987.[10]

The refinery is part of an aging national oil transportation system that includes oil trains and pipelines. In ExxonMobil's case, pipes connect the refinery to the port complex less than ten miles away (receiving crude, delivering marine fuel, though both had slowed because of port congestion). Other pipelines deliver aviation fuel to Los Angeles International Airport, push gasoline to distribution terminals in the region, and receive the world's thickest, densest crude from oil fields 160 miles north in the San Joaquin Valley. The explosion halted production at this massive petroleum nexus and contributed to a dollar-a-gallon spike in prices at the pump within weeks. A state investigation later brushed aside the oil giant's assurances about safety and found nineteen separate violations at the plant, six of them deemed "willful." The state determined that ExxonMobil had failed to fix the fluid catalytic converter that exploded despite knowing for nine years that it was faulty. The defect made the explosive buildup of pressure undetectable until it was too late. Investigators reported that the explosion could have been catastrophic, with a risk that a cloud of highly toxic, potentially fatal hydrofluoric acid could have been released into populated areas. The $560,000 fine levied against the oil giant was nothing compared to the consequences to the region's consumers and businesses, who saw the results at the pump. While the rest of the country was paying $2.43 a gallon in the fall of 2015, Southern Californians were paying an average of $3.45. As the investigations continued, Exxon abruptly announced it had sold the troubled refinery to New Jersey–based PBF Energy for $537.5 million.

Meanwhile, near Los Angeles International Airport—long despised for being accessible only by car while being surrounded by jammed highways and streets—excavation of a new subway line kicked into high gear today. When it's completed by 2020, after a quarter century of failed plans and pleas to make airport travel easier, Angelenos will finally be able to do what other major airport passengers have long taken for granted: they will be able to take the train to the plane. Well, almost. The new eight-mile line will fall just under a mile short of the terminals, requiring some sort of people mover to close the last leg of the trip—a barrier of money and inconvenience that could easily slow the project and deter riders until that last mile is built.

The people mover is promised as part of a separate $5 billion modernization and makeover for the city-owned Los Angeles International Airport, which could be completed as soon as 2024, if only to sweeten the city's bid to host the Olympics that same year. LAX, one of the busiest and most reviled airports in the world, was originally built in the sixties and expanded in the eighties (for the last Olympics held in LA), which means it was designed to accommodate comfortably about half the 70 million travelers a year who now elbow their way to the gates. The double-decked horseshoe-shaped access road that provides the only way into the nine separate airline terminals is a nightmarish, paralyzed, roaring mess of cars, buses, shuttle vans, and fumes virtually every hour of the day and night. It is by far the most hated mile drivers face anywhere in Los Angeles. The modernization plan calls for killing the horseshoe and directing incoming traffic to remote parking lots served by the vital, but still mythical, people mover.

In other air travel moves this day, the Federal Aviation Administration conducted a final round of debugging and shakeout tests for its $40 billion NextGen air-traffic control system. The long-awaited, long-delayed, and over-budget system is supposed to increase safety and efficiency for air travelers weary of delays and canceled flights. NextGen, which would pass its tests and be up and running by March 2015, would replace the aged computer system known simply as "Host," which had run plane traffic in U.S. airspace with floppy-disk-era technology for forty years. Stage one of the rollout had NextGen taking over at FAA's twenty main air control centers that run high-altitude plane traffic nationwide. The next two stages—upgrading 1970s-era analog radio systems and replacing airport ground radar centers with satellite-based GPS—is supposed to happen by 2020. The old ground radars still in use are safe but horrendously inefficient and slow to update, which means air traffic controllers have to keep aircraft separated by large buffer zones to avoid collisions. With the new satellite systems, the 7,000 passenger planes zipping over America at any given time should be able to fly much closer together without compromising safety, which will speed up takeoffs and landings. And that—in theory—will not only make the skies more friendly to fly again, but is expected to save the economy an estimated $22 billion a year by reducing fuel waste, chronic congestion, and flight delays.

Finally, on February 13, U.S. transportation secretary Anthony Foxx kicked off a national bus "infrastructure tour" today, with the long-shot goal of garnering support for his $478 billion federal transportation plan, which focuses on what he calls the "beyond traffic" approach. By that he means to seize on the lessons of Carmageddon and emphasize *mobility* over car culture: less sprawl and highway expansion, in favor of policies promoting "close-in" growth connected to existing cityscapes and transit options. "The old approaches aren't working," the affable former mayor of Charlotte, North Carolina, proclaimed. To illustrate his point, his bus crawled through a South Carolina interstate highway confluence so confusing and in such disrepair that local commuters have dubbed it "Malfunction Junction." There are hundreds of such "spaghetti interchanges" around the nation, Foxx complained, but no money for the fix.

The goal of Foxx's tour was to pitch his idea to replace the federal tax on gasoline. That tax is supposed to fully fund the national Highway Trust Fund, but it falls hopelessly short. For strictly political reasons, the 18.4-cent federal gas tax, which doesn't begin to pay for the system's needs, is not likely to be raised anytime soon, given that it's been frozen since 1993. So Foxx and the Obama administration proposed to replace it with a new 14 percent tax on foreign corporate earnings, which could only be spent on transportation and that would generate double the current gas tax revenues. On paper, this would be a tax reduction for American businesses: the current tax rate for foreign profits is 35 percent. However, companies generally avoid paying any of that by parking their cash offshore. Foxx's proposal would close that loophole and fully fund transportation in America in the process. A number of major corporations might support this. Apple Inc., for example, had by 2015 nearly $200 billion in cash accumulated offshore from foreign sales that it would like to use domestically for a variety of projects, including buying back a large block of its own stock. At a 35 percent tax rate, repatriation of that money is a nonstarter; under 15 percent, and the company just might bring the cash home. Apple's CEO testified before a Senate committee in 2013 that he was looking for just such a deal so long as it was "fair to all taxpayers."

Nevertheless, Congress would shoot down Foxx's plan. Americans just don't like to pay for their roads, and members of Congress like asking them to pay even less. And what does all this have to do with the pizza delivered to your door?

Everything. We couldn't be more contradictory about this: More than nine out of ten American voters believe it's important to improve the country's transportation infrastructure, and eight out of ten say it's vital in order for America to stay competitive with other nations. Yet seven out of ten voters adamantly oppose raising the federal gas tax from its 1993 levels.[11] Which is why Congress is basically cooking the books with accounting gimmicks to keep the system afloat year to year, deferring critical repairs and modernization projects year after year.

The face of modern transportation is built on equal doses of wonder and fear. There's the wonder of a dinosaur-size boring machine chewing subway tunnels with whirling steel teeth beneath a bustling cityscape. Or the near-instant gratification of a morning click-and-buy of that new camera or shirt or case of food for your very expensive diabetic cat, and finding your purchase on the doorstep that afternoon. Then there's the counterbalancing frustration of street-level gridlock that leaves you in a cold sweat as you crawl toward the airport, your flight time fast approaching. Or those hideous interstate merges with the exit ramps placed exactly at the rushing, turbulent confluence of several freeways, forcing you to make a rapid series of white-knuckle swerves across multiple crowded lanes of traffic just to get home. That's the transportation world 320 million Americans see and experience every day—delightful one moment, a misery the next.

But that's just the visible surface of movement and mobility. Beneath lies the hidden *heart* of our door-to-door system of systems: the constant, never-ending war against overload. In every facet of its past design and future spending, in all its rails, wheels, water, wire, and pipe, the real transportation battle and rationale are all about the overload.

This was true in the age of canals, when the nation's first true goods movement system was built out by 1840, boasting 3,000 miles of waterways linking east to west, with a continuous water route all the way from New York to New Orleans. And it was true of the bigger, faster, more capacious Iron Horse that overtook the canals with its greater speed and reach, and of the new highways and ports and sprawling logistics centers that came next. Sooner or later something new and more capable always gets added to the transportation mix to augment or supplant the old, easing the overload. While some iterations fade fast—the "golden age" of dirigibles was barely an eye blink—other innovations endure for generations, sometimes far longer than logic and balance sheets suggest they should.

Transportation choices are never purely rational ones. Habit, myth, culture, and lifestyle can and do trump practicality and efficiency in transportation as much as in matters of fashion or art—as demonstrated by America's most popular passenger vehicle, the pickup truck. Developed a century ago for farmers, builders, and other craftsmen, it began as a humble utilitarian alternative to the horse-drawn cart. Now, in an age when few Americans farm and most live in cities, the pickup truck is predominately a lifestyle choice playing on the urban cowboy mythos. Pickup truck popularity today is largely unrelated to most owners' job requirements, making them a statement rather than a need, and it is yet another force pushing overload because of the pickup's inherent inefficiency compared to standard passenger cars. The once homely Ford F series of pickups, America's most popular vehicle of any kind, boasted three no-frills editions—the Heavy Duty, the Contractor, and the Farm & Ranch Special. Now it comes in models the salesmen refer to among themselves as "Cowboy Cadillacs," with features such as handcrafted eucalyptus wood interior trim; leather seats that heat, cool, and massage driver and passengers; and sticker prices of $60,000 and up.

If a new technology, structure, or method works well or delights its users or both—if it saves time and money or boosts self-image—then users flock to it. The elevated wooden bicycle highways built in Los Angeles in 1900 with a 10-cent toll—and unfinished routes that left riders stranded—got a well-deserved thumbs-down and closed after five years. The New York City subway, in continuous service since 1904, got a big thumbs-up when 150,000 riders boarded the very first day for a nickel apiece and never stopped coming. It's the if-you-build-it-they-will-come, induced-demand process at work. Making it well, and then later making it bigger or better, promotes demand, a principle that holds true both inside and outside the transportation space. Edison found this with his lightbulb, Ford with the car (and more recently the pickup truck), IBM with the home computer, Apple with the smartphone. The new and strange can quickly become widespread necessity—or at least perceived as such.

Once that happens, once the harvest can be moved by canal barge instead of wagon, once the shiny new streetcar can zip so much faster from suburb to downtown, or that paved road is laid or that new dock is built, demand for what they offer soars. Overload recedes for a time. But with success comes the threat of new overload. Sooner or later there are too many boats, too long a wait for the trolleys, too many cars clogging the street, and capacity has to be increased. There has to be another lane, a bigger dock, a larger vessel, new traffic signals to smooth the flow while easing the gridlock and carnage. Adding capacity takes planning, time, and money and, once built, the increased capacity induces even more demand. The cycle feeds upon itself, the Carmageddon experience repeated time and again, until finally capacity can no longer be expanded. Then either overload reigns, or something new, something more capacious, bursts on the scene to create a system reset: the horse-pulled trolleys of old, supplanted by more capable cable cars, superseded by electric streetcars, surpassed by gasoline-powered private cars filling wider streets and broadened highways to capacity, which will be surpassed by . . . no one is sure. The time is long past for yet another system reset, and still we wait.

Much of this uncertainty stems from the feast-or-famine transportation investment pattern the country has fallen into during the past century. America has seen only two big transportation building booms during the last hundred years—which is to say, there have been two mammoth, nationwide expansions of capacity. That's it. That's what we're living on.

First came the landmark infrastructure projects that modernized America and put the country back to work during the Great Depression. There were the Lincoln Tunnel, the Triborough Bridge, and LaGuardia Airport in New York. There was the Overseas Highway—the name is literal—that connects Miami to Key West with its famous Seven Mile Bridge. This first boom produced the Hoover Dam, the Grand Coulee Dam, the Golden Gate Bridge, rural electrification, and America's first freeway (in Los Angeles, of course, the Arroyo Seco Parkway, also known as the Pasadena Freeway). This boom also produced the dual massive water transport systems known as the Colorado River Aqueduct and the Central Valley Project that supply, respectively, the faucets of arid LA and Southern California, and the farms of California's Great Central Valley that supply much of the nation's produce and all of Domino's Pizza's sauce (and the rest of the world's pizza sauce as well). Most of the nation's almonds, artichokes, plums, celery, carrots, broccoli, cauliflower—the list is immense—are sustained by this water transport system, enabling the Central Valley to generate 8 percent of the nation's agriculture (by dollar value) on less than 1 percent of its farmland. Many other iconic working monuments of transportation that serve as economic engines to this day were built by "make-work" Depression-era programs such as the Works Progress Administration, along with three quarters of a million miles in more mundane but no less essential roads, more than 7,000 bridges, a thousand airstrips, and tens of thousands of nameless dams, levees, tunnels, and drainage projects.

The second big wave of transportation investment took place two to three decades later, during the fifties and sixties, when most of the modern Interstate Highway System was built, capitalizing on (and then feeding) America's unprecedented postwar prosperity and growth. This 48,760-mile highway network has most visibly transformed passenger car travel, but the real revolution lies in what it did for trucking. The interstates paid for themselves many times over by allowing the unprecedented explosion in goods movement that, in turn, enabled America's modern, import-driven consumer economy to take off and take root. The rapid growth of ports and containerization—and such retail giants as Walmart and Amazon—needed the interstates in place to make their business models practical and complete, because existing rail freight could not handle it all or serve the last legs of the goods journey from warehouses to stores and homes. When the interstate system celebrated its fortieth anniversary in 1996, it was said to have returned by then an estimated $6 in economic productivity for every dollar spent building it.[12] That number has easily doubled since then.

Public investment in American transportation has never again achieved anything close to those two big door-to-door building booms. Since the completion of the Interstate Highway System, America has spent about half as much on its infrastructure as some European nations, calculated as a share of gross domestic product. China invests almost five times as much as the U.S. Sweden, which is far less dependent on cars than America due to its well-used and well-loved mass transit systems, still spends about as much each year on roads as the U.S. (again, not in absolute dollars, but as a share of GDP).[13] Despite the widespread perception that American taxpayers overspend on their transportation systems, they have rarely been more miserly than they are in the twenty-first century.

Some of the most important elements of America's transportation system—all those iconic structures we drive and ship and fly through, under and over—are now fifty to one hundred years old. They face problems of age, decay, obsolescence, or insufficient capacity—and deferred maintenance. Sixty-five percent of roads are rated in less than good condition. One in four bridges require major repairs or are too weak for modern traffic. About 45 percent of Americans have no access to mass transit. The World Economic Forum ranks America's infrastructure sixteenth in the world in overall quality, with the rail system coming in fifteenth and road quality sixteenth.[14] The country that built the first transcontinental railroads and the first interstate highway system has indeed fallen on hard times.

Meanwhile the U.S. Department of Transportation estimated in 2015 that economic losses just in the form of wasted time and fuel because of traffic jams have reached $121 billion a year. This is the aged hardware—4 million miles of roads, streets, rails, waterways, and pipes—with which the United States moves $20 trillion in goods, the stuff of our daily lives, from state to state, city to city, and country to country every year. The material moved through all those systems weighs in at more than 17 billion tons.[15] By way of comparison, the Empire State Building weighs 365,000 tons. America moves goods equivalent to 46,575 Empire State Buildings door to door every year. If all that was loaded on just standard 53-foot semitrailers, it would require 425 million big rigs to move it, with every truck filled to the legal 80,000-pound limit. That would take about eighty times more trucks than the entire U.S. fleet of registered semitrailers.

The cost of decades of neglected maintenance and modernization was demonstrated in July 2015 when the eastbound side of an obsolete bridge on Interstate 10 near Hell, California (you just can't make this stuff up), collapsed during heavy rains. Interstate 10 is a major route eastward for goods coming out of California ports, farms, and factories, with more than eight thousand big-rig trips a day passing through that stretch of highway. Alternative routes slow consumer, commercial, and industrial goods deliveries by up to a day, which in today's economy is a body blow; the extra fuel and lost time alone from this one bridge accident cost the trucking industry an estimated $2.5 million a day. And that's just one little bridge on one truck route. Multiply those losses by the 61,000 other bridges in as bad or worse condition nationwide, and the risk to commerce and prosperity is almost incalculable.[16]

The poor condition of America's transportation infrastructure leaves us less prepared today for severe weather than at any time since World War II. Severe wind, rain, and flooding puts added pressure on poorly maintained and unsafe bridges, overpasses, tunnels, dams, culverts and roads. It took only one big storm—Hurricane Sandy in 2012—to devastate the New York and New Jersey region and paralyze coastal areas up and down the East Coast, disrupting the systems that move our food, water, energy, products, and people door to door. Sandy inflicted $75 billion in damages, and a return to normalcy required many months. If the predictions of climate scientists are borne out, we are entering an era when super storms such as Sandy may become all too common. If so, America should be building more resilient transportation systems and structures as soon as possible, rather than allowing them to grow weaker through deferred and unfunded repairs and maintenance.

Beyond roads, passenger rail infrastructure is decaying, much of it ancient. A pair of century-old, leaky, and unstable tunnels handle all passenger trains crossing the Hudson River between New York and New Jersey, one of the busiest—and most urgently in need of repair—rail tunnels in the country. A plan to build new ones was scuttled by the governor of New Jersey in 2010 so he could channel the funds into road projects.[17] Half the 1,000 bridges in the Northeast Corridor—the nation's most critical and crowded rail lines, running from Washington D.C., to Boston, with stops at all the major cities in between—are about a century old. Several tunnels still in use were dug just after the end of the Civil War. Ridership on Amtrak in the corridor hit 11.6 million in 2014, a 50 percent increase since 1998, but the system is crumbling beneath those passengers, and lags far behind the rest of the world in both speed and reach.[18]

Century-old water mains lie beneath the streets of American cities, that crucial but hidden transportation system that grows more fragile every year. The University of California, Los Angeles, campus and parts of affluent West LA were flooded in one epic break in July 2014, when the principal area water supply line burst, spouting a geyser thirty feet high and spilling 75,000 gallons per minute for nearly four hours during the height of a statewide drought. Other communities, from Flint to Chicago to Toledo to the nation's capital, have made residents ill with aged and contaminated water supply lines; the American Society of Civil Engineers' annual Report Card for America's Infrastructure found that America's water supply systems had reached the end of their useful lives in most locations, grading the system with a D. The underground pipelines delivering oil and gas have suffered similar catastrophes. And the electrical grid that transports our power is sustained by hardware so old, fragile, and lacking in backup systems that a Pentagon commission concluded in 2011 that six well-placed rifle shots could take down the entire East Coast grid serving more than a hundred million people. A similar attack could cripple the West Coast, not only blacking out the grid but also shutting down most of Los Angeles's water supply.

No special access or phony ID would be required to accomplish such epic sabotage of America's power transportation system. Just as ominous: the nearest company that could manufacture the obsolete, truck-size replacement parts for those damaged in such an attack is located where all our stuff seems to come from—China. Estimated manufacturing and delivery time via cargo vessel would be four months.

America in the twenty-first century is uniquely cursed: nearly every major method of moving people, resources, energy, and goods faces overload. The lone exception is the 140,000-mile national freight rail system, America's door-to-door superstar. Its unique status as a system run by privately funded virtual monopolies helps explain how it is better able to cope with overload. Deregulation in 1980 allowed the freight rail companies to abandon less profitable lines, including almost all passenger service, to set their own rates, and to expand capacity only where there is a clear return on investment. To be fair, that investment has been substantial—$575 billion since 1980 and $29 billion in 2015 alone[19]—but freight rail does not offer a model that can be replicated throughout the transportation system of systems. Government has no such flexibility to close costly major roads, water mains, or bridges that serve whole communities, or to charge higher user fees at will so that roads can pay for themselves like rail freight. It's akin to the difference between private schools that pick and choose students and set their own tuitions, and public schools charged with educating everyone with no tuition at all. The White House, the Congress, the state transportation bureaucracies, the city transportation departments, and the advocates for every mode from bus to bike to car to feet agree there is an urgent need to face up to the resulting overload that infects every other method of transport. But as yet no clear vision of what comes next has emerged in America's door-to-door conversation beyond adding a lane here or a bypass there—the traffic equivalent of going on a diet by loosening your belt.

My Lyft driver, Tom, is talking about the string of cargo vessels visible up and down the coast, all waiting to berth, each piled high with containers like floating ziggurats. The line extends twenty miles south of the port, all the way down the coast to Huntington Beach, forming a grim backdrop for the surfers and beach walkers. What should have been a source of well-paid work for Tom has turned into a need to moonlight as a rideshare driver.

Of all the transportation headlines and behind-the-scenes developments, the port slowdown has dominated the news of Friday the thirteenth, as it had for many days before, this crippling backlog at this most important hub in the goods-movement universe.

With the port so backed up, farmers with citrus to ship to Asia are watching their produce rot, or they're forced to sell domestically for far less than cost. Beef exports are down, as are nuts and alfalfa. Total lost farm sales are running about $1.75 billion a month. Target stores can't get their normal supply of Easter baskets. Apparel maker Nature USA is running out of yarn. Tommy Bahama had to stock last season's shirts instead of the new fashions in transit from Asia. Levi Strauss is missing a third of its products. Softline Home Fashions, a leading importer of home decor, can't get its new line of curtains to its main customers: Walmart, JCPenney, and Bed Bath & Beyond. They all have payrolls to make, investors to satisfy, bills to pay. They need their products.

And then there's the disaster of Valentine's Day. Tomorrow.

"My Valentine's business is destroyed and my Easter season is on the verge of being destroyed!" laments the CEO of Megatoys, a Los Angeles–area business that normally hires six hundred seasonal workers to assemble Valentine's Day gift baskets.[20] But now his eighty shipping containers of hearts and toys and all the trappings of prepackaged romance are stuck as sea. As are his competitors'. They are all within sight of the port—some are even berthed—but they cannot get unloaded. They might all as well still be in China. No products means no jobs for those workers, no baskets for the sweethearts, no profits for Megatoys.

Guys like Tom, banking on their lucrative longshoremen's salaries, now can't pay their mortgages with far smaller earnings from Lyft and Uber. For all the media celebration of the sharing economy and the "gig economy," the pay is low and the benefits nonexistent. The regional and national economies could suffer long-term damage as some shippers have threatened to shift to ports in Mexico or to traverse the Panama Canal and go with East Coast ports only too happy to scoop up the business. The system that always "just worked" is not supposed to creak and stutter like this, and the usual power brokers and lobbyists haven't helped.

In desperation, the White House dispatched the secretary of labor to Los Angeles to broker a deal—or, as my Lyft driver Tom puts it, to "knock some heads" before the damage to jobs and the economy escalates from a Valentine's Day massacre to a full-blown recession restart.

Only later, when this intervention actually works and a settlement is reached before the end of February, does it become clear that the labor dispute was not the main problem, and that the other factors overloading the system could not be so easily solved.

"It's a great job, working here," Tom says of the port. "But who knows what the future will bring? How do we handle the overload?"

Chapter 7

THE LADIES OF LOGISTICS

Who rules the seas? Does the power lie with the great national navies and their mighty warships? This is certainly where the United States is the undisputed big dog. America has the impressive firepower of the U.S. Navy, which also bears an even more impressive cost: $168 billion a year to keep the biggest guns in the water on standby, just in case of threat or attack.[1]

Or should we look to a different type of sea power, the arena of cargo ships? Their purpose is not intimidating enemy targets but actually stocking Target stores, along with every other retailer, business, and home in America. Along with all their thousands of other customers, those cargo ships just happen to deliver 80 percent of the components the U.S. Navy and the rest of the American military relies upon. The Pentagon outsources as much as everyone else. When it comes to the superpowers of global shipping, the U.S. barely ranks as a bit player.

In a concentration of power unlike any other sector of the transportation system, six steamship companies, none of them American, control more than half the goods in the world.[2] Twenty global companies—most of which have joined forces in four immense ship-sharing alliances—control almost every product traded on earth.

This has been the quietest conquest and surrender in world history, one in which the entire United States happily and somewhat obliviously participated because consumers love above all else low prices at the cash register, and there is no question that globalization has delivered that part brilliantly. The miracle of modern logistics and ultra-efficient global transportation technology has made achieving those low prices possible, although beneath the gleaming tech lies the crudest of foundations. All it took was two things: divesting America of its once-mighty cargo fleets and shipyards; and outsourcing a major chunk of consumer goods manufacturing to countries with pay, benefits, environmental practices, standards of living, and working conditions that would never in a million years be tolerated on American soil. The thrill of the checkout-line bargain masks the reality that Americans pay elsewhere for those low prices in the form of shuttered U.S. factories, lower wages, a shrinking middle class, a growing inability to pay for roads and bridges, massive public subsidies of the health and environmental costs of transportation pollution, and a nation—including its armed forces—that can no longer function without massive amounts of Chinese imports shipped aboard Korean-built vessels owned and operated by foreign conglomerates.

Of the six cargo powers that control a majority of global goods movement, Denmark-based Maersk Lines is the leader, at the top in numbers of ships, in cargo capacity, in revenues, in profits, and in constructing the biggest and most advanced cargo ships in the world. Maersk (with subsidiaries in oil platforms, oil drilling, trucking, and port terminal operations) handles nearly 16 percent of the world's cargo all on its own. That's the equivalent of one out of every six items for sale at Walmart, making the globe-spanning Danish shipping company the undisputed world superpower of ocean commerce, based in a country with a population slightly less than the state of Maryland's. And if that's not enough, Maersk has partnered in a mega ship-sharing alliance with the Geneva-based Mediterranean Shipping Company—the world's second biggest container ship line. Together the two companies' "2M Alliance" control a combined fleet of 1,119 vessels capable of hauling 29 percent of the world's goods.[3]

Not a single missile, cannon, or gun bristles from this container ship fleet. Yet, when this shipper—or any one of the other handful of powerful alliances—wants something, entire transportation systems fall over themselves to accommodate. Billion-dollar projects are launched. Port terminals are torn down and rebuilt. Freeways are expanded, railroads rerouted, trucking companies meekly accept backbreaking new costs and delays—all because the shipping lines are so good at playing country against country and state against state, always ready to move their business to a more obliging port or partner if they don't get what they want. Is Los Angeles not giving shippers such a good deal? Well, there's always Houston or Savannah, Mexico or Canada. It's been that way for decades now, the hidden price tag on those inexpensive goods America once made onshore but now outsources abroad: 97 percent of our clothing, 98 percent of our shoes,[4] two-thirds of our home furniture,[5] most consumer electronics, toys and bikes; they're all everyday products, and they all come from abroad. There's a reason why the global container fleet's cargo capacity has jumped from 11 million tons in 1980 to 169 million tons in 2010[6]—because the growing consumer-driven economy demanded it and then became dependent upon it.

Seventy percent of America's gross domestic product takes the form of consumer spending,[7] much to the delight of those big shipping lines that connect Chinese factories to American stores and closets. The big steamship lines, both metaphorically and quite literally, have been driving the train for decades, one of the greatest forces, if not *the* greatest, propelling the transportation system of systems.

Still, the shipping domain is not entirely a one-way street. The shipping superpower rule book was rewritten—at least a bit—when a most unlikely port director took over the gritty docks of Los Angeles. Unlike the shipping industry insiders who usually take charge of big ports, this director was a marine biologist with a vintage pink Thunderbird, a penchant for baking, and little regard for the glass ceiling that made her one of only two female directors of the nation's eighty-five major ports. Before she left her post in 2014, she would take on the shippers, the truckers, and every other entrenched constituency at the nation's biggest port complex. On her watch, the dirtiest harbor with the dirtiest ships—a major source of air pollution in America's smog capital, Southern California—would transform into an international model of greener transportation even as her port continued to be number one on the continent. She would become a charter member of an unofficial group of transportation movers and shakers, the Ladies of Logistics—the LOLs, as they call themselves. The LOLs not only buck the tide in a boys'-club industry but their influence demonstrates that, for all the impersonal immensity built into the global transportation machinery, a few brilliant or dedicated or determined individuals can have an outsized impact on how we move ourselves, our stuff, and our world. And, before her term was up, she would see the current congestion crisis coming, she would plan for it, and her legacy just might help find a cure.

"Good times" is how Geraldine Knatz recalls her eight years at the helm of one of America's most vital transportation hubs—the most desirable and difficult job in the port business. "Not bad considering my first job here was to study life in the port waters back in the seventies—which wasn't easy, because there was *no* life in the port waters back in the seventies. I suppose I've come full circle. And so has the port."

Tuna, not containers—that's what the Port of Los Angeles was most famous for back when Geraldine Knatz's forty-year career at the twin-ports complex began. The evolution that followed is a microcosm of the transformation of the entire global supply chain as local industry dominance gave way to offshore supremacy, canned tuna included. This was long before the container revolution created a Southern California mega-port dominated by superpower shipping lines. Fishing fleets and canneries were the shot callers back then in the Los Angeles harbor—sardines in the early twentieth century, led by the French Sardine Company, then tuna once the sardines were depleted and French Sardine rebranded itself as Star-Kist.

Imports were a relatively small piece of the port puzzle even when Knatz arrived in the mid-seventies for her graduate studies in marine biology at the University of Southern California. At the time, Star-Kist and its cat food brand, 9Lives—along with Chicken of the Sea, Bumble Bee, and every other tuna enterprise in the country—operated out of Los Angeles, which had become the biggest tuna production site in the world after World War II. The fishing fleet and canneries employing thousands occupied the wharf space now filled by container ships and terminals. Just by the smell on the morning breeze, the locals in San Pedro overlooking the harbor could tell if it was a human food day or a pet food day. That distinctive aroma of an open sack of dry dog or cat food would permeate the waterfront on pet days, distinct from the briny, heady fish scent of regular tuna canning. And no matter the day, human or pet, the canneries dumped their waste from fishing, cooking, and canning into the harbor as they had done for more than half a century. It was into this miasma that young Geraldine and her wetsuit were quite literally thrown, as the port launched a belated effort to cleanse its polluted waters.

Paddling around in cannery waste had not been her dream job. She grew up watching the prime-time TV specials starring Jacques Cousteau aboard his intrepid research vessel, the *Calypso*, as he played with the dolphins, spoke rhapsodically in his melodic French accent of the shy intelligence of the gentle octopus, and advocated for ocean preservation. The scuba-diving adventurer's broadcasts inspired thousands of baby boomers to pursue a career in the ocean sciences, Knatz among them. She had grown up in the northern New Jersey township of Wayne, where her father worked as a factory die cutter and her mother kept the house, both of them forced by the Great Depression to take jobs as teenagers rather than finish high school. They maintained a hard-times work ethic all their lives, requiring their daughter to get a job and start paying room and board at age sixteen in addition to her high school studies. She didn't dare complain, as she had been given a better deal than her brother, who had to start earning his keep at age fourteen. But it did drive her to covet a career she could love rather than merely endure.

In her spare time, young Geraldine collected pond samples while walking home from school or work, using the microscope her brother had received one Christmas to identify the protozoans within. While at college, she landed a research internship studying port pollution near Fort Hancock, New Jersey. The National Oceanic and Atmospheric Administration (NOAA) scientists working on the project at the army base took pity on her when she confessed that, on the way there, she had to pretend she dropped her quarter on the ground at every other toll both on the Garden State Parkway. That way she would be waved through without paying, a tactic she adopted because she could afford to pay only every other quarter. Her supervisors ultimately let her move into a disused old barracks at the base to avoid the tolls, then finally gave her a paid position. At night she used to lie on the beach and watch the big break-bulk cargo ships go around the Sandy Hook point en route from New York Harbor. It was the beginning of her fascination with cargo ships and ports.

After graduating from Rutgers University and winning admission to USC's postgraduate program, she drove to California and found a little apartment in the Silver Lake area of town, not far from Dodger Stadium—a fashionable, eclectic neighborhood now, unfashionable and very affordable back then. She left for class or the harbor early every morning, exchanging greetings with her elderly neighbor, Maria, who spoke to her with a thick Russian accent. Only after Maria died did Knatz learn she had been living next door to Rasputin's daughter. The unusual and the strange often seem to follow in Geraldine Knatz's wake. (The root meaning of her Germanic surname, she says, is "troublemaker.") One of her employers, she recalls, made the entire port staff undergo handwriting analysis to reveal their personality traits, an expression of this particular port director's fondness for fringe science and psychic phenomenon. She thought it was a bit weird at the time, but as she was promoted immediately after the analysis came back, she never complained. That director, a local businessman whose expertise was not in ports but in amusement parks and hotels, left after a mere year in office. The most important legacy of his tenure would turn out to be Knatz's promotion.

Knatz began her long career at the port as a student researcher attached to a water-quality project funded by the tuna canneries. The canneries hoped that science could show that their dumping of fish waste into the harbor was actually a beneficial "bio-enhancement." It wasn't. The harbor had been stripped of life and the oxygen that supports life, and the canneries were a big part of the problem.

"We'd go out and if we found a little bit of oxygen in the water we'd celebrate," Knatz recalls. "Back then, there was basically nothing living there. The water was dead."

But it was not irredeemable. Knatz spent the next three years researching and diving with USC as she earned her master's degree, then took a job as an environmental specialist at the Port of Los Angeles while pursuing her doctorate. Her work then focused primarily on improving the abysmal water quality in the harbor. This was not altruism or good corporate citizenship at work. Air pollution, water pollution, and extinction threats to such signature species as the national symbol, the bald eagle, had become so grave that a Democratic Congress passed and Republican president Richard Nixon signed a series of sweeping, powerful, and publicly popular bipartisan environmental laws in this era, among them the Clean Water Act of 1972. Under the law, canneries had to start treating their waste and dispose of it properly. Leaking oil and fuel infrastructure had to be repaired or replaced throughout the port. One of Knatz's first tasks was to dive the port to assess the damage caused by leaking oil after a docked tanker ship exploded. She dove with a bunch of white sticks in hand, measuring the toxic gunk that had accumulated on the harbor bottom by plunging the rods into the sea bed like dipsticks into a car engine.

It would take more than a decade's work to show substantial improvements in the port waters, to nurture kelp beds transplanted shoot by shoot and rock by rock (the underwater plants cling to rocks, not sand), and let nature work its slow magic. Slowly, the harbor waters once again became infused with life-giving oxygen and the ecosystem that had always belonged there revived. Life returned to the harbor, not least because the commercial tuna industry collapsed. Overfishing, the lure of cheap labor in American Samoa and Asia, and the ebb and flow of the new global economy led the big tuna brands to flee one by one, with all gone by the end of the eighties except for Chicken of the Sea, which hung on until 2001, then closed what was the last tuna cannery in the continental U.S. Thailand-based Thai Union Frozen Products now owns Chicken of the Sea while British private equity firm Lion Capital LLP owns the Bumble Bee tuna brand—yet more once-proud domestic products outsourced, with many more thousands of miles added to America's cupboards in the process. On average, each American household consumes nineteen cans of tuna a year.[8]

Huge export container terminals eventually rose where the canneries once stood, but smaller-scale commercial fishing enterprises flourished anew around the harbor as water quality improved and overfishing ended. The work by Knatz and her colleagues exceeded all expectations.

But its success notwithstanding, her environmental work offered Knatz little opportunity for career advancement. The port brass perceived her department less as an asset and more as a hindrance to the port's main mission: shipping, commerce, and growth. It was her job to question projects and their impact on the environment, and to suggest ways of reducing the harm. The projects ports like to do most—dredging, filling, and creating more capacity for exhaust-spewing ships, cranes, and trucks—*always* have a negative impact. And reducing those negatives usually imposes additional cost. Never mind that Knatz's work insulated the port from even more costly lawsuits and delays.

She would have happily moved into a different department at the port with more responsibility, earnings potential, and advancement possibilities, but Knatz soon learned there were few opportunities then for women in the core lines of business at the port, where the workforce was overwhelmingly male. She felt the prevailing attitude back then was that she was already fortunate to be the only female in a "nontraditional" job at the port, meaning she didn't work a clerical, secretarial, personnel, or public relations position. When she applied for other jobs—in marketing, for example—she recalls being turned down because, a manager explained, she couldn't play golf with "the guys." Or because the wives of the men on the staff wouldn't like it if she went with them on foreign trade missions. Some of the cultures they had to visit abroad to drum up business for the port would think she was brought along only "for sport," one supervisor suggested, which confused Knatz until she realized he was referring to sex. In another time or with another employee, this could have been the stuff of very costly lawsuits, but Knatz recalls never considering such a move. "I'm more of a nose-to-the-grindstone type of person, and figure if I let the work speak for itself, I'll prevail."

In 1981, Knatz accepted an offer to run the environmental department at the rival Port of Long Beach. Although there is of necessity some collegiality and partnership on broad issues and projects that affect two gigantic ports that happened to share a single harbor, they have for the most part been locked in fierce competition for customers ever since the late 1920s. That's when the discovery of copious oil reserves beneath the previously ritzy Long Beach neighborhood of Signal Hill financed the creation of a second commercial port in San Pedro Bay, luring both Ford and Procter & Gamble to build massive factories nearby. When Knatz came over to Long Beach six decades later, the Long Beach port was considering a range of large building and improvement projects to accommodate the shift to container ships and the approaching tsunami of Asian import goods. Knatz's challenge was to deal with the environmental impact of those projects, a seemingly impossible task. The 3,000-acre port eventually would be jammed with ten mammoth piers, eighty deep water berths, dozens of miles of on-dock railroad track, and twenty-two cargo terminals—six of them sprawling container yards, six for bulk commodities (salt, gypsum, cement, petroleum coke), five for petroleum and other bulk liquids, and five break bulk terminals for such cargo as cars, lumber, steel, aluminum, and iron ore. There's no way to accommodate all that without environmental harm, and no room to spare to set aside sufficient wildlife preserves, parks, or buffer zones that might, as the law demands, "mitigate" the damage to the environment.

The difficulty of striking an agreeable balance between profit, planet, human health, and technical feasibility is one reason why so many big, landscape-altering projects—particularly huge, emission-spewing transportation projects—get held up for years and so often end up in court. State and federal laws unambiguously require mitigation, but the question of how much is always contentious, with project boosters vying to do the least possible, and communities that might suffer from an expanded port, freight yard, or freeway pushing for the most possible. With the port caught in the middle, Knatz hit on what was then a daring plan, although it's now accepted practice: she wanted to do environmental mitigation *outside* the port for work *inside* the port. She began using port funds to restore nearby Pacific coast wetlands, build public parks and beach facilities for communities near the port, and clean up landscapes and habitats near but not in the port that had far higher ecological value than the waters of a bustling commercial harbor.

The regulatory bureaucracy initially bristled at the unorthodox approach, but local communities loved it and began lining up to pitch their own projects as candidates. The tide turned completely in favor of this "outside-the-docks" approach when both Long Beach and Los Angeles ports agreed to provide a combined $60 million for the restoration of the critical Bolsa Chica wetlands twenty miles south of the harbor. This important breeding ground for migratory birds and other marine species had been badly damaged by oil drilling, roadbuilding, and encroaching development. Today it's an ecological preserve and permanent conservancy—a slice of nature initially undone by the rapidly expanding transportation system, only to be brought back from near destruction by even more transportation expansion at the ports.

In 1988, Knatz left her job running the Long Beach port's environmental department to become director of planning instead, which finally put her where she wanted to be: in the thick of the ports' core business. There she would push through a long list of expansion projects, most notably a $2.4 billion mega-project known as the Alameda Corridor. This twenty-mile cargo expressway for trains links the twin ports to the transcontinental rail lines that pass through downtown Los Angeles. By the time it was done, the Alameda Corridor would rank among the biggest and most successful public works projects ever undertaken in the U.S., a public-private partnership that, unlike most mega-projects in recent years, finished on time and on budget.[9] It is the envy of other American cities that lack such a connector between port and transcontinental rail system. Chief among them is New York City, which is number one in the U.S. when it comes to dependence on trucks for goods movement, and so suffers from unusually intense diesel truck emissions. Of the more than 400 million tons of freight that move through New York annually, 90 percent moves by truck.[10] Nationally, trucks move about 70 percent of freight.[11]

The Alameda Corridor project solved that problem on the West Coast. It came about because the economic boon of rapid growth in container traffic and imports spawned a Newtonian reaction, equal and opposite, bringing a swarm of trucks that boosted congestion and smog on the major streets and highways connecting the port to the rest of the world. Traffic engineers and local planners, who predicted worse ahead as the port continued to grow, believed creating an efficient method of moving more cargo by train instead of by truck offered a solution.

The key word here was "efficient," for trains were nothing new at the port. There had been a rail connection between downtown and the port in place for more than a century—indeed, for almost as long as Los Angeles had anything more than mudflats and a few nailed boards to call a port. Knatz, so fascinated by the history of the port that she would write numerous articles and two books on some of the more unusual aspects of its past, dug into this end of the story gleefully. A visionary entrepreneur and self-made millionaire by the name of Phineas Banning built the first railroad between the port and the warehouse districts of downtown Los Angeles all the way back in 1869. Banning also masterminded the first dredging and wharf construction in San Pedro Bay. Previously, ships had to anchor off the coast and shuttle goods ashore on small boats. At the time, Banning's railroad traversed open coastal country, sparsely settled and otherwise difficult to cross. His Los Angeles & San Pedro Railroad represented a huge transportation advance from wagons banging over rutted roads and range. The new transportation technology began the transformation of a muddy backwater port into a thriving West Coast economic engine.

A few years later, the Southern Pacific Railroad bullied itself into ownership of Banning's achievement, as its monopolist tycoon leader, Collis P. Huntington, essentially extorted its sale in exchange for connecting the city to the transcontinental railroad line he was building. Huntington also demanded the city pay his railroad the then princely sum of $600,000 ($11.5 million in 2014 dollars) as well as gift Southern Pacific with sixty acres of choice downtown LA real estate. Otherwise, he vowed the Southern Pacific, originating in San Francisco and moving south down the coast before chugging eastward, would bypass

Los Angeles entirely. And this pre-Hollywood Los Angeles of the 1870s, a mere village of 5,000 or so, could simply wither and die without a rail connection. The big rail lines were the nineteenth-century lifeblood of commerce, travel, and migration, as the ruthless Huntington knew all too well. Capitulation by the city and Banning was inevitable, but that was just the start of Huntington's machinations. Next he launched a campaign to strangle the very port in San Pedro harbor he was supposed to be serving by imposing horrifically high freight rates on the railroad Banning built. Under the new rate schedule, it often cost more to move goods the twenty miles between downtown and the port than it cost to ship the same goods to Hong Kong. Huntington wanted to force customers to take their import and export business to his own pet port project in the beach town of Santa Monica thirty miles up the coast, where he had had constructed what was then the longest wharf in the world. That's where he wanted the leading port of the West Coast to take root: on his land, his wharf, his town. His goal was control in Southern California of nothing less than the two dominant transportation modes of the day, marine and rail, thereby owning Los Angeles's future all by himself.

Huntington's bold bid to create what would have become the greatest trade and transport monopoly in U.S. history failed after a years-long battle known as the "Free Port Fight," but only after his foes in Congress just barely thwarted his clout and money. They succeeded in designating San Pedro Bay over Santa Monica as the site for an all-important breakwater construction project by the Army Corps of Engineers, who were charged with creating Southern California's first protected calm-water harbor. Construction of that breakwater was such a big deal at the time that a special wire was set up to connect the White House directly to the San Pedro waterfront so that President McKinley could push an electric button on April 26, 1899, signaling the construction crews to commence work. Boulders were carried from inland quarries by steam locomotives pulling long lines of flat cars piled high with the rock needed to build the massive breakwater, a dike to calm the tides and protect docked vessels from storms and high seas. Like so many other vital pieces of the transportation system of systems, the design and location of the future hub of commerce on the West Coast—and the very existence of Los Angeles as a major city—rested not on some grand municipal study, engineering analysis, or the project's objective merit. The biggest port on the continent owes its existence to the vagaries inherent in a few powerful individuals' struggle for wealth, dominance, and the big win, along with a president's giddy desire to throw the switch on what was at that moment the nation's largest public works project. Had the question been limited to simply choosing the most naturally ideal place for a major Southern California port, neither location in Los Angeles would have been chosen. If geographic merit alone were the only consideration, McKinley's magical button would have connected to San Diego and that city, rather than Los Angeles, would have grown up to be the future hub of West Coast cargo, commerce, and population growth. Whether it's the selection of a port location, or a twenty-six-lane freeway expansion in Texas that makes traffic worse, or LA's decision after World War II to dismantle the most extensive light rail and streetcar system in the country (then, a half century later, spend billions and decades trying to undo that folly), America's transportation future owes as much to greed, gamesmanship, and hubris as sensible design. Perhaps more.

When the dust settled in the war of the would-be ports, Los Angeles stood connected to the Eastern U.S. and the rest of the world by rail and sea, and the village of 5,000 of the 1870s grew to a city of 100,000 by 1900. Just ten years later it would be home to 300,000. The rural outskirts between downtown LA and the port began to evolve into the dense checkerboard of communities, industry, immense rail yards, and the grid of streets and roads that it is today. But during this urbanization process, what had been a breakthrough railroad route to the port through the hinterland became an antiquated roadblock and a continual frustration for both shippers and drivers. Motorists found themselves stacked up at railroad crossings twenty cars deep while waiting for excruciatingly long and slow freight trains to pass. By the 1980s, Banning's old railroad right of way consisted of four low-speed branch lines owned by the Southern Pacific Railroad with more than two hundred street-level crossings. That's an average of ten intersections of car and train traffic every mile. The route had become such a jigsaw puzzle that freight trains rarely averaged speeds of more than 10 miles an hour, 15 on a very good day. On bad days, the twenty-mile trip from port to downtown took four hours. A bicycle rider, even a jogger, could beat that time. These trains ran up to 6,000 feet in length—a mile and an eighth—which meant drivers trapped at those crossings typically waited twenty to thirty minutes each time a train passed. Those trains ran continuously throughout the day. This was a mess, a system pieced together across more than a century, becoming increasingly dysfunctional as the city grew up around it. The rail line to the port was not designed for modern traffic, much less the high risk of death and destruction whenever human drivers are asked to sit patiently for twenty minutes with nothing more than flashing lights barring them from crossing the tracks.[12]

Solving this traffic miasma with a single rail mega-corridor that darts above or below street traffic—an expressway for trains—may sound obvious in retrospect, but formidable barriers deterred it for many years. There was cost to consider, and property rights, community fears, the engineering challenge of building an immense trench in a dense urban area, and the sheer "It's the way it's always been" inertia that keeps bad transportation systems in place for decades. The railroad had lucrative monopoly control of the right of way and no incentive to ease the pain to the populace or the rest of the goods movement system by investing billions—which is why the idea originated in government, at a regional planning agency where a transportation visionary by the name of Gill Hicks worked.

One of Knatz's first big hires was Hicks, whom Knatz stole away to spearhead the project. The port could do what the private sector could not accomplish on its own. Although it's a publicly owned asset, the Long Beach port receives no funds from the city but runs off its earnings, where it acts as landlord to all the businesses that operate there. Every decision has to make business sense at the port, and investing in a better rail connection—though a financial nonstarter for Southern Pacific—would be a customer magnet for the port, drawing cargo with the promise of far faster goods movement. Hicks would lead the project and go on to be the director of the corridor program after it opened for business as its own quasi-public agency. Knatz, meanwhile, partnered with her old employer, the Port of Los Angeles, becoming the chief negotiator for acquiring the Southern Pacific rail lines, stations, yards, and rights of way. This ended up being a hard-fought $235 million deal for control of the corridor that the bullying Collis Huntington got for free through threats and intimidation. Still, the deal transformed goods movement at both ports while relieving one of the worst traffic-jam generators on that end of Los Angeles.

Completed in 2002, the Alameda Corridor project isolated the rail lines from street traffic with a series of bridges, underpasses, overpasses, and the centerpiece of the project, a ten-mile trench three stories deep and fifty feet wide, a man-made canyon for trains. This enabled freight trains to zip downtown from the port in a half hour or less instead of the two hours plus it used to take, without slowing car traffic a single minute. Some of the trains run nearly two miles long, but there are no flashing lights and no crossing gates. On average, each train hauling cargo out of the port takes three hundred trucks off the road, and there are nearly fifty such trains a day. That's 15,000 truck trips a day removed from crowded freeways and smoggy air.[13] About a third of the goods coming into the twin ports now moves out through the Alameda Corridor rail system. Without it, the growth of cargo flowing into the port would have had a crippling impact on freeway traffic and smog in the region long ago.

In 1999, Knatz was elevated from planning director to the number two job at the Port of Long Beach: managing director. She thought she might be in line for the top job eventually, but when a leadership change left her future uncertain, Knatz sought to return to her old employer, the Port of Los Angeles, when a new mayoral administration was voted in and the top job at the port opened up. The application process devolved into the sort of odd turn of events that seems to crop up wherever Knatz is involved. It started when she received a form-letter rejection in the opening round of the port job search. This seemed odd, given her long career and relationships on both sides of the harbor, not to mention her position as second in command at America's number two port. But she was prepared to shrug it off and move on. Her husband, however, insisted that the letter had to be a mistake. Knatz loved him for his faith in her, but she said there was no way the top headhunting firm hired to recruit a new port director would be so bumbling. She couldn't call back and say, *Gee, did you really mean to reject me?* How foolish would that look? When he wouldn't relent, Knatz finally placated her husband by reluctantly placing the call, only to be flabbergasted when, after a long silence, the headhunter exclaimed, "We sent you the wrong letter!" Then Knatz learned she was already scheduled for an interview at an airport hotel that Saturday evening—the last one scheduled.

When she arrived for the one-on-one, however, the apparently exhausted headhunter dozed off in the middle of Knatz's interview. Mortified, she returned home and complained, "There I was at a hotel on Saturday night with a man who's not my husband. And he falls asleep!"

The ridiculous interview, like her errant rejection letter, turned out to be an unreliable indicator of Knatz's job prospects. She soon learned she had made it to the next round, then on to a meeting with the mayor. The next thing she knew, Knatz had been tapped to run the most important port in America.

It turns out her background as a biologist and environmentalist left her uniquely qualified for the times. The port complex was under siege as the worst air polluter in California. Virtually every major project to improve the LA docks, the terminals, and the surrounding infrastructure sat frozen by lawsuits and protests. Her predecessor had worked for two decades for steamship line giant Maersk before coming to lead the port, an insider status perfectly suited to dealing with the shipping industry. But he was perceived as insensitive to the concerns of the surrounding community by such organizations as the Natural Resources Defense Council, which had targeted the port as a leading threat to health and environment. Knatz, who had come to the port as the scientist-outsider so many years before, was uniquely positioned to settle the skirmishes with environmentalists and surrounding communities and get the port modernization back on track before overload paralyzed operations. The new mayor made the new port director's mission very clear, she'd later say: Fix it.

Knatz, in turn, told her staff they would be adopting a new in-house motto to describe goals to simultaneously grow the port with massive new infrastructure while radically cutting pollution at the same time: "It's impossible. It's difficult. It's done." This had been the progression on the Alameda Corridor project, she said, and the same approach would fix the port. At first these apparently conflicting goals would seem impossible, she said. Then achieving them would seem very, very difficult.

"And then we'll be done."

The environmentalists' claims were not wrong. The port had been identified by state scientists as the major source of air pollution in the region. The ships were a big part of the problem.

There are about 6,000 container ships in the world (out of a total of 90,000 cargo vessels of all types). These behemoths move 120 million container-loads a year, worth $4 trillion.[14] They are also prodigious consumers of one of the dirtiest fossil fuels on the planet. The growth of maritime shipping and the giant container ship has powered an unprecedented explosion in world trade, but also an environmental disaster.

Bunker fuel, it's called: the cheapest, dirtiest form in common use is up to 1,800 times more polluting than the diesel fuel used in buses and big rigs,[15] and little more than a waste product left over after everything else useful is extracted from crude oil. It has the consistency of asphalt; a person can walk on it when it's cool. The big cargo ships burn so much bunker fuel that they don't measure consumption in gallons but in *metric tons per hour*, with the really big ships consuming two hundred to four hundred tons a day. One large container ship burning this type of fuel spews out more sulfur and nitrogen oxides—the precursors of smog and particulate pollution, as well as a major contributor to the ocean acidification that threatens fisheries and coral reefs—than 500,000 big-rig trucks or roughly 7.5 million passenger cars.[16] That means just 160 of the 6,000 such megaships in service today pump out the same amount of these pollutants as all the cars in the world. New regulations that took effect in 2015 compel shippers to use a cleaner, less polluting type of bunker fuel when they cross the two-hundred-mile limit at the edge of U.S. waters, which has reduced ship emissions close to shore. But the cleaner bunker fuel is more expensive, and so many shippers hover just outside the limit when waiting for a berth opening at their next port of call, then race in at the last possible minute, thereby maximizing their use of the dirtiest, cheapest fuel.

The cargo fleet is also a prodigious source of carbon emissions—about 2 to 3 percent of the global total.[17] Although that's only between a third and a fifth of the global-warming gases emitted by the world's cars,[18] it's still a big greenhouse gas footprint for such a relatively small number of vessels. If the shipping industry were a country, it would be in the top ten drivers of climate change, and its billion tons of carbon dioxide and equivalents put it ahead of Germany, the world's fourth largest economy. At current rates of growth, the shipping industry that hauls 90 percent of the world's goods will be two and a half times its current size by 2050; absent a serious effort to become more energy efficient, it could be generating a staggering 18 percent of global greenhouse gases by then.[19]

Through a very deceptive accounting loophole, none of these big ship emissions "belong" to any one country. They happen in international waters for the most part, and so for the purpose of calculating the greenhouse gas emissions of nations, they simply don't exist—on paper. They very much exist in terms of their impact on climate, oceans, and health. As the world's biggest consumer, America is the beneficiary of a substantial portion of container shipping, as well as the manufacturing of the products on board. The United States is the per capita world leader in global greenhouse gas emissions, and second only to China in absolute volume of emissions. Much has been made in the U.S. Congress and press of China's overall lead in greenhouse gas emissions, but about half of the growth in Chinese carbon since 1990 is the result of offshoring and globalization—China's explosive growth as a manufacturer of products for export, with the U.S. the biggest customer. Those emissions, in other words, are as much ours as China's. We haven't outsourced just jobs and manufacturing; we've outsourced our carbon. Even with that accounting trick making the U.S. appear greener than it really is, the average American produces about twice the carbon emissions of the average European, and 3.5 times the average Chinese citizen.

When these ships approach the port and dock, the practice in the past was to keep the engines running as they burn through tons of bunker fuel to maintain power aboard the ship. The bunker fuel emissions made working at the port unhealthy and polluted neighboring communities.

Within five years of Knatz's arrival, pollution at the port dropped so dramatically that even the most ardent critics were dishing out praise: emissions, particularly those related to diesel and ship engine exhaust, were down as much as 76 percent even as container traffic increased.[20]

Several big changes in the handling of ships and cargo led to this startling turnaround, which initially made the twin ports of LA and Long Beach outliers, then world leaders in a growing green ports movement.

The Port of Los Angeles became the first U.S. commercial port to install cold ironing for docked container vessels. "Cold ironing" is the industry term for the use of giant power plugs to connect ships to the shoreside electrical grid while docked. This allows them to shut down their engines (thus the "cold" part of the term, which harkens back to the days of red-hot coal-fired engines) and still keep their essential electrical and refrigeration systems operating without spewing fumes and toxins.

Work on the very first shore power plugs in LA started before Knatz's tenure began, as part of a settlement of a massive lawsuit filed by the Natural Resources Defense Council that had halted construction at the port for years. But the new port director turned that grudging acquiescence by a predecessor into a port-wide asset, touting zero-emissions ship power as a solution to the single biggest pollution source on the coast. The shipping industry initially objected to the cost of the new shipboard equipment, but Knatz argued that they'd benefit from a combination of fuel savings and community goodwill—and the likelihood that an investment in clean tech and good corporate citizenship on this point could make future projects the shipping industry desired more palatable to the public.

The Port of Long Beach undertook similar reforms. By the end of 2014, half the cargo ship calls at the ports were supposed to be with engines off and shore-based power engaged. That goal was not quite achieved—at Los Angeles, it was closer to 35 percent, as the shipping lines stalled on installing the necessary equipment. Even so, the pollution reductions from ship exhausts were dramatically reduced, exceeding goals that were not required for another three years.[21]

Other U.S ports at first tried to capitalize by positioning themselves as cheaper, easier, and less regulated alternatives, and a few shippers did shift some cargo. But even with the new green requirements, few ports in the U.S. could match LA or Long Beach on the time and costs of shipping. Economics and geography ensure this: to ship a container from Asia to America via Los Angeles or Long Beach takes twelve days on average at a cost of about $1,800. It's another five days' transit time by rail or truck to

Chicago or the East Coast, a total transit time of seventeen days. By comparison, bypassing the West Coast in favor of an all-water journey from Asia to the East Coast via the Panama Canal takes twenty-seven days at an average cost of $4,200 per container, not counting the time and expense of shipping from the port to its final destination. Shipping via the Suez Canal takes even longer: thirty-seven days.[22] As long as the twin ports kept modernizing and preparing for larger ships and container loads, the shipping lines would not abandon them over a green initiative that other ports around the world had begun to copy anyway.

Under Knatz's regime, the Port of Los Angeles spent an average of a million dollars a day on maintenance, modernization, and upgrades. She likens a port to a bridge, like the George Washington or the Golden Gate, which must be painted regularly to protect the metal structures from corrosion but which are so huge and take so long to finish that, once a paint job is completed, it must start again almost immediately at the other end of the span. As part of this effort at upkeep, Knatz gradually swapped out old and dirty diesel-powered dockside machinery for handling and moving cargo in favor of motors that rely on natural gas, batteries, or other lower-emissions technology.

The third and most contentious measure the twin ports initiated during Knatz's time in office was a clean-trucks program that phased out older, heavily polluting big rigs that served the ports as drayage vehicles. Dray trucks—the term dates back to horse-drawn, open-sided carts for hauling goods short distances—fetch containers at the port and move them to nearby warehouses, rail yards, and other intermodal terminals, where the container goods are unpacked, sorted, repacked, and handed over to longer-range shippers for transport to the actual cargo owners. Drayage is an essential but troubled leg in the consumer supply chain, with the drivers often paid and treated poorly, working as contractors rather than employees, many of them owner-operators driving older—and dirtier—vehicles.

Trucking and retail trade associations tried to block the clean-trucks program, but their lobbying and lawsuits failed to halt the phasing out of old dirty trucks. By 2012, no trucks from before the 2007 model year could operate at the port, and the diesel emissions throughout the area measurably lessened.

Other players in the port ecosystem were more supportive than the drayage companies and truckers of the changes. The shipping lines, whose single greatest expense is fuel, were looking for new efficiencies anyway, and cold-ironing could be a money saver for them in the long term. The terminal operators who receive the goods coming off the ships were sick of lawsuits over environmental concerns halting their expansion plans and saw the green-ports initiative as a means of relieving the bottleneck—an investment that would pay dividends by allaying community concerns and allowing more growth. The longshoremen's union and local communities were pleased because they saw direct improvements in working and living conditions. Neighborhoods near the ports worried less about respiratory disease and elevated rates of childhood asthma. Environmental critics praised the transformation. Dockworkers experienced the greatest relief. Many had spent their careers working in choking fumes, their clothes black with diesel particulate at the end of the day. Crane operators, hovering over the ship's exhaust stacks as they lifted out cargo, had inhaled the stuff all day long. "They want to earn a good living," the union president told Knatz at the outset of the green program. "They do not want to pay with their lives for a stronger economy."

There were concerns about the cost of going green in the short term, but these were balanced against long-term gains in fuel savings and lower maintenance costs. Instead of driving business off to other ports, as was initially feared might occur, the green programs won international praise and inspired similar reforms worldwide.

There were other threats to the ports and, by extension, the whole goods-movement system, against which Knatz knew any tussles over the green-ports initiatives paled by comparison. The port and its transportation connections to the rest of the world were facing overload with a public that neither understood nor felt sympathy for the port's needs. Modern Los Angeles has always styled itself as the headquarters for beaches, surfing, movies, and television—the capital of car culture, the center of sprawl, the home of Hollywood. But the host city to the biggest seaport complex on the continent, the spout through which so much of the consumer economy pours, has never viewed itself as a port town. In short, Knatz had to get the message out: another project of the magnitude of the Alameda Corridor was needed—and probably more than one—and for that she needed allies. That's when she turned to the LOLs.

The fifty or so members of the Ladies of Logistics call themselves a social group. They're women at work in the male-dominated logistics world who get together for power walks and potlucks and Geraldine Knatz's famous bake-offs—and for networking.

Transportation, so often referred to as a "system," is dizzyingly fragmented. Its leaders, workers, and innovators occupy their own silos, laying their plans and setting them in motion for bigger vessels or new distribution centers or expanded freight yards all according to their own individual needs, often with little or no input or consideration of other elements in the supply chain. But when a group of LOLs gathers, they bring together top transportation executives across all modes and industries:

rails, roads, ships, retail, academia, real estate. Among them are developers and owners of warehouses and distribution centers; members of the California Transportation Commission and the National Freight Advisory Committee; the vice president of the Pacific Merchant Shipping Association; a leader from the longshoremen's union; the head of the Coast Guard's Los Angeles command; leaders of chambers of commerce; and other major players in ports and government.

When the vice president for government affairs at the Burlington Northern & Santa Fe Railway faced protests and lawsuits against a proposed new freight yard, dubbed the Southern California International Gateway, her fellow LOLs—Knatz and Elizabeth Warren, who leads a harbor business association called FuturePorts—spoke out publicly to support the plan. It would be a benefit to the entire regional transportation system, these Ladies of Logistics argued to a skeptical public, easing congestion by replacing trucks on the freeway with boxcars on rails, lowering pollution, making goods movement—and therefore the goods themselves—cheaper for all. Later, when a port terminal expansion plan Knatz coveted aroused neighborhood concerns about traffic and property values, the LOLs from local universities appeared with data showing economic growth and jobs in the community tied directly to growth at the port.

Fran Inman, senior vice president of Majestic Realty and appointee to the National Freight Advisory Committee, brings to Washington the views of all the LOLs. Her own perspective is unique: Inman's company is the largest private industrial real estate owner in the U.S. and warehouse landlord to such diverse tenants as Amazon, General Electric, In-N-Out Burger, Walmart, and Crayola. She found that many people are fond of saying the transportation system is broken, but the real problem is that it's invisible. "When it's working, which is most of the time, no one pays attention," she complained at one LOLs event. "People tune in only when something's not working."

As she became immersed in transportation policy, Inman was flabbergasted to learn that America has no national goods-movement strategy. Knatz told her other countries are surging ahead of the U.S. on this score, particularly in marketing their seaports as alternatives to Los Angeles and Long Beach. "Canada has an entire national freight strategy worked out," she said. "And their strategy is to steal my business."

The LOLs cut across all parts of the transportation system, and because many of them have projects that require public approval, they spend a good deal of their time *explaining* to politicians and public why ports and trains and all the myriad projects that advance or maintain movement matter. As a result, for a "social group," the LOLs have amassed considerable clout. It's hard to go to a major transportation meeting or conference without bumping into a few of them—or hearing them mentioned and thanked by name by legislators or congressional committee chairmen, who had at one time or another turned to the LOLs to find the right data or the right experts or to drum up support for a highway project or port expansion or rail safety bill.

Their work has led to the near completion of a unified freight policy at the national and California state level—a means of prioritizing (and paying for) improvements that will have the most impact. They also were instrumental in the launch of Knatz's last big project at the port before leaving her post[23]: revitalization of the Port of Los Angeles's waterfront as a center of dining, shopping, and commerce, as it was in decades past. The plan includes Knatz's legacy project, the $500 million port business and scientific research center called AltaSea. She believes this project may finally convince Los Angeles that it really is a port town and generate the innovations needed to solve the opposing problems of insufficient capacity and too much environmental impact. Knatz would have liked to have stayed on a few years longer to work on this under the new mayor, she says. But now her successors—Gene Seroka, the former shipping line executive now running the Port of Los Angeles, and Jon Slangerup, who ran Federal Express Canada before taking over the Port of Long Beach—are grappling with the overload caused by megaships and their mega-alliances on one side, while on the other, lawsuits and public opposition to building more rail yards and freeway capacity for freight moving in and out of the port. Communities that will have to live near any new infrastructure—and cope with the fumes, noise, and health effects—insist that they will withdraw their opposition only if the new trucks and trains, as well as the powerful machines and vehicles that load them, use clean, low-, and zero-emissions technology. The ports themselves are already making such a shift, but the private rail and trucking companies responsible for the logistics outside the ports' gates say they can only clean up slowly because of the cost. And so the impasse continues.

Being part of the LOLs, hearing the diverse concerns of port workers, railroad leaders, activists, and shipping companies made clear to Knatz why the door-to-door machinery that ought to function as a seamless system so frequently succumbs to paralysis instead of building consensus when it comes to reducing overload. The absence of an overarching national strategy combines with lack of public understanding to create failure after failure. Traffic lane expansions that don't fix car traffic are celebrated, funded, and built, even as highway connectors and new rail gateways that would reduce truck congestion and pollution are opposed and languish for decades. Auto safety recalls that save hundreds of lives on the road receive the highest priority, but fundamentally flawed system designs that kill tens of thousands of Americans every year are blandly accepted as the cost of doing business. Massive warehouse builders are courted and given tax breaks by communities eager for jobs and growth, and then those same communities complain when truck traffic and smog triples overnight. The costs of these disconnects have been ignored and papered over for decades, the LOLs say, but they are now reaching a critical mass. And nowhere is this more visible than at the ports.

There are the new, larger ships with cargo capability beyond even the massive Port of Los Angeles's current capacity to handle—requiring yet more expansion and construction. There are the shipping line alliances that save the steamship companies money by pooling resources but throw everything else out of kilter because of the crazy mix of cargo stacked together on single ships. It's as if Federal Express and UPS tossed all their stuff on a single truck to save money, then brought it all to a U.S. Postal Service warehouse to sort, where havoc predictably ensues. These changes have saved shipping lines fuel and money, but they have also created backlogs and congestion at ports worldwide, with port directors caught in the middle. Suddenly the natural allies of shipper and port are at odds, with no easy solution in sight. The rest of the transportation system has to suck it up and try to solve the problem the shipping lines created.

Then there is the melting Arctic that is beginning to alter long-established trade routes by opening once ice-clogged routes at the top of the world. In a few years, these new routes will send even more cargo streaming into the West Coast ports. Finally, there is the constant need to modernize and automate the movement of cargo from ship to shore and to and from the port, set against the need to keep peace with labor forces that can make or break a port's rankings in the world.

Preparing for these threats and disruptions requires something else: getting public buy-in. Knatz had to grapple with public indifference and ignorance of how transportation works and why it matters—and how there's more to the door-to-door world than long commutes and rush hour traffic. She realized just how fundamental and dangerous a problem she and every other port director faced early in her tenure when, during a hearing on the proposed expansion of a port terminal, a member of the public objected on the grounds that the project would generate more truck trips on already crowded freeways.

"Why do I need a port?" the woman asked. "I have Walmart."

There were murmurs of agreement from an audience that felt that these trucks were indeed deplorable, their presence a barrier to the travels and commerce of "real people." They had no idea that the two things were so intimately connected, Knatz realized then: how the shoes that woman was wearing were almost certainly made in China. Or that a majority of shoes from China moved through her port. Or that there was an entire web of connections, from a rail trans-loading terminal south of Los Angeles, to a freight yard outside Kansas City, to a truck depot in the heartland, all of which were absolutely critical in maintaining a world where an annoyed consumer can say, "I have Walmart." Indeed, it was this woman's consumer choices—and those of a couple hundred million more Americans like her—that drew the ships and trucks to the port in the first place. That woman, Knatz felt like blurting, wasn't the victim. She was the cause.

"People just have no idea," she said later. "They don't make the connection."

Chapter 8

ANGELS GATE

Nearly nine out of ten Americans say their daily commute begins with the crank and cough of a car motor. The remainder report that their morning journeys start with the music of coins falling into fare boxes, or the well-oiled click of bicycle freewheels, or the ancient rhythm of soles on sidewalks. The U.S. Census keeps careful track of these commuter statistics. And, data aside, daily experience and preferences suggest these numbers make sense.

Yet the numbers mislead.

For all our preoccupation with its foibles and frustrations, our actual morning rides or walks to work are the least part of our commutes. They are not the first steps in our daily movements but the very last legs in a far longer, far more mysterious daily journey built on the thousands, even millions of miles embedded in our everyday lives, choices, and actions.

So where does our commute truly begin if not in our driveways?

More than anyplace else, the starting line can be found high atop the windswept Pacific bluff of Angels Gate where, seven days a week, the most valuable shopping list in America is created.

There, inside the whitewashed, antennae-studded headquarters of the Marine Exchange, a very pleasant, very busy mother of four by the name of Debbie Chavez crafts the Magna Carta of the buy-it-now, same-day-delivery world: the Master Queuing List. With it, Chavez holds the lion's share of America's consumer economy in her hands.

If you drive it, wear it, eat it, buy it, drink it, talk into it, type on it, or listen to it, some portion has first passed by Angels Gate. From the morning cup of coffee to the tires on your car to the bike you bought to replace that car to the shoes on your feet and the smartphone in your pocket, all or part passed in and out of the control of Debbie Chavez before it entered your life.

"We do keep busy here," Chavez observes with the casual understatement of a woman who does something extraordinary so often that she mistakes it for routine.

For generations, Angels Gate has been prized for its coastal vantage point. First came the cannons placed early in the last century to stave off unwanted invasion via the waters below. Later Angels Gate became the ideal spot to track a more benevolent but no less disruptive commercial invasion in the form of a daily, miles-long procession of giant cargo vessels laden with . . . everything.

The blocky Marine Exchange control center's one notable architectural feature, its ocean-facing picture window, offers one of the great juxtapositions in California topography. To the right is a million-dollar view of sun-dappled waves, the rocky enclave of the Palos Verdes peninsula, and the gorgeous green mountain of Catalina Island in the distance. To the left is a $400 *billion* view of the hard metal angles and industrial bristle of the twin Ports of Los Angeles and Long Beach, the busiest harbor complex in North America, serving as both barometer and driver of the U.S. economy.

Each day in predawn darkness, Chavez and her crew of marine information specialists arrive at Angels Gate to chart the approaching parade of cargo vessels, gathering cryptic information received via phone, e-mail, and old-school fax from the world's far-flung maritime shipping lines. The product of these labors is a master daily schedule for a hundred or more impending ship departures, arrivals, crossings of the two-hundred-mile international limit, and shifts to the marine terminal docks from remote harbor anchoring spots (the waterfront equivalent of the doctor's waiting room). Once dockside, the mammoth ships need two to five days to unload and reload before leaving for their next port of call and making room for the next vessel, which means every berth has a waiting line behind it.

Those lines are no small matter: one container ship can be as long as four football fields laid end to end, with as many as 7,000 giant shipping containers filling its hold and stacked on its deck. (There are even bigger container ships sailing now and larger still on the drawing board, but the Southern California ports lack the necessary colossal cranes and other infrastructure to accommodate them—for the moment.) At any one time, fifty such ships may be lingering at the dual ports of LA, waiting their turn.

As much as a billion and a half dollars' worth of product passes through these twin ports every day. Delays can break businesses nationwide and cost consumers millions, interrupting the cornucopia of endeavors the port makes possible: the installation of new air bags in recalled cars, the delivery of the latest computers, the flow and pricing of gasoline, the supply of those all-important silicon chips that are inside everything—from coffeemakers to cash registers to the controllers that keep our traffic lights functioning at hundreds of thousands of intersections. Without that cargo and its timely delivery, all that and so much more grinds to a halt. The Master Queuing List is the essential first step that sets off a well-choreographed transportation chain reaction—the commute to end all commutes.

First, Debbie Chavez sends out the list to inform the work of the traffic controllers and Coast Guard officers at the Marine

Exchange "Watch" peering at their radar and computer displays. They direct and police the approaching vessels.

Then the Master Queuing List is used to schedule the port pilots who race out to meet the ships and guide the laden behemoths in and out of their berths.

The list is next used to staff the day shift with the right number of crane operators, those princes of the docks who lift twenty-ton containers from impossibly tight quarters with the finesse (and pay scale) of brain surgeons.

Then comes the assembly of longshore gangs to unload the goods, and the stevedores in the marine terminals who move and prepare the cargo for shipment out of the port.

Finally, the Master Queuing List is used to dispatch the 40,000 or more big-rig truck trips that swarm into, out of, and around the twin ports every twenty-four hours, carrying the cargo out into the concentric circles of warehouse distribution centers, freight depots, and rail yards that make up America's goods-movement ecology.

Other ports in the American West, East, and South have their own lists and vital roles to play, but LA's Master Queuing List holds a unique place in the trade and transportation continuum because so much of what Americans buy and sell comes from the Pacific Rim, and the twin ports ringing San Pedro harbor are Asia's leading point of entry.

A third of U.S.-bound consumer goods, and far higher percentages of some, pass by the Marine Exchange. That makes Angels Gate and Debbie Chavez the one essential stop for *everyone's* commute—long before you even leave the house.

The complex ballet required to move a product, any product, from door to door—and the overload that affects and infects that dance—begins most often at a port. In the U.S., the leaders for containerized goods are the twin ports of Los Angeles and Long Beach, the Port of New York/Newark, and Savannah, Georgia, followed by Seattle, Virginia, Houston, and Oakland. Their history is the story of transportation overload in microcosm, particularly the Port of Los Angeles, which has become both a cause of system-wide overload and one of its greatest victims.

In the late nineteenth and early twentieth centuries, the first cargo ships sailing for San Pedro Harbor on the south side of LA had two options. Their captains could throw up their hands and bypass Los Angeles entirely in favor of the then thriving, now vanished port of Anaheim Landing to the south. Or they could anchor outside the mudflats of San Pedro, unload goods onto small boats, and then row them ashore to waiting horse-drawn wagons. Exports could repeat the process in reverse. It was not a delightful way of doing business.

As soon as the village of Los Angeles showed the least inclination to grow, the inefficiencies and limitations of this infrastructure-free method became intolerable and the demand for more capacity irresistible. Small piers were built in the natural harbor beneath the Palos Verdes Peninsula. This induced a few more ships to come calling on San Pedro with bigger loads, and soon larger piers had to be constructed. Commerce grew further, as did the opportunity for profit, which is when the enterprising Delaware native who founded the new California port town of Wilmington, Phineas Banning, began the first of many efforts to dredge harbor channels to accommodate even larger ships seeking to do business in LA. This is when Banning also began laying plans, then rails, to link the port with downtown by train.

Soon the demands of commerce in a rapidly growing city overloaded those accommodations, too, and the simple piers evolved into massive wooden wharves built with a graceful curve so the longest portion of the docks paralleled the coastline, enabling far longer piers with more berths without jutting far into the shipping lanes. Then, when the cargo moving across those extended wharves became too heavy to move by hand or horse, train tracks were laid on the piers, with steam engines pulling boxcars laden with goods and grains to and from the loading zones. Those mechanized wharves drew even more ships, creating a shortage of protected warehouse space, yet another problem of overload to solve. In 1917, the prospering city responded by building Municipal Warehouse No. 1, a six-story concrete behemoth, then the largest warehouse west of Chicago and the most visible seaside structure on the West Coast for many years. Approaching ships could navigate by sighting that tall squarish building before anything else at the port was visible. It quickly became a center of activity and commerce in California, an object of fascination for its size and the port's engine of growth for decades, a pinnacle of induced demand.

Three thousand piles were driven into the bedrock for Municipal Warehouse No. 1's foundation in order to support the 27,000 cubic yards of cement and 1,200 tons of reinforcing steel. The three-and-a-half-foot-thick walls yielded a building with a controlled climate year-round without air-conditioning. The warehouse cost $475,000 ($9.5 million in 2014 dollars) to build, with twelve acres of storage space and sixteen electric hoists serving those six floors. The ground floor boasted the killer feature: multiple rail lines running through the center in a recessed bed with concrete platforms on either side—a subway station for freight, with room for twenty-four boxcars unloaded entirely inside the warehouse.

For most of the twentieth century, Municipal Warehouse No. 1 served as the only bonded warehouse at the port, the landmark where everything from gold bullion to circus elephants to the rail coach that carried Winston Churchill to his grave ended up after arriving at the port. In the 1950s, the interior of this singular warehouse was designated as its own foreign trade zone for duty-free imports and re-exports, which made stepping inside the building the legal equivalent of leaving the country.

More than any other facility at the port, this warehouse and its unrivaled capacity put Los Angeles on the map as a center for global trade, relieving overload for years while accommodating a growing port and a growing city. Only the rise of containerization put an end to Municipal Warehouse No. 1's dominant role in cargo commerce on the West Coast, and although its retro-cool interior is used to this day for storage, its most prized role now is as a convenient set for Hollywood filmmakers. The ground-floor railroad tracks and concrete platforms are fairly easily dressed up as a convincing New York City subway station, close enough to every major Hollywood studio to limit crew costs to day rates, no overnights required. (The port is regularly used for various film backgrounds; the Mediterranean-style hacienda and palm tree–studded grounds of the Coast Guard commander's house on Terminal Island is another favorite, figuring prominently in a number of movies, including standing in for Guantánamo Bay, Cuba, in *A Few Good Men*.)

Even after the rise of open-air container yards over bulk warehouses, the giant building's height kept it in use for many more years to meet another sort of rising demand at the port: the demand for information. The warehouse's high roof and vantage served as the first and most enduring home for the Marine Exchange and its all-important Master Queuing List.

This critical starting point in the goods-movement system evolved with the same pattern of demand and overload as every other major component in the door-to-door world. In the early 1900s, messenger boys were posted on hilltops overlooking the port, where they could watch out for approaching cargo ships. They'd try to identify incoming ships by the names marked on their hulls or, failing that, by the rigging of sailing vessels or by markings on the smokestacks of steamships. Once a ship was spotted and at least tentatively identified, the runners would then race back to the shipping agents who employed them, who then would compete with one another to dispatch pilots, tugboat captains, and stevedores to meet the new arrival. Running boys were soon augmented with men on horseback to speed the process.

By 1920 the inefficiency of having dozens of runners and riders racing (and sabotaging) one another on the hilltops near the port, then showing up at the office mere moments before a ship arrived, became untenable. A shipping agent and local customhouse broker came up with the solution when he founded the precursor of the modern Marine Exchange, which he called the Maritime Exchange and Sailing Club. W. H. Wickersham got permission to station lookouts atop the best dockside vantage point available, the roof of Municipal Warehouse No. 1, closer and better than any hilltop. His spotters, all experienced mariners—many of them retired Navy and Coast Guard signalmen—had telescopes on the rooftop to spot the incoming vessels. Once they made a sighting, they had flags and semaphore lights for communicating with the ship crews, and megaphones to shout the information down to coworkers on the ground, who would spread the word to all the port players onshore to prepare for a new arrival. As the ships passed close by, the lookouts would shout berth instructions through the megaphones to the crew.

The enterprising Wickersham supplied the arrival information to the maritime industry for a fee, and although his service proved popular, his dream of finding fortune in the lookout business fell short. After two years of no profit, he decided to shut down his rooftop crow's nest, but was persuaded instead to turn the business over to the Los Angeles Chamber of Commerce. The chamber would subsidize the operation because of its obvious benefits to the business community and the local economy.[1] Renamed the Marine Exchange of Los Angeles and freed from the pressure of turning a profit, the service became a twenty-four-hour, seven-day-a-week operation. Onshore communications were switched from megaphones to telephones, with the tower lookouts relaying their information to a central office, which then assembled a daily report of the LA shipping news. The exchange posted the information in the building lobby in San Pedro, but the real breakthrough came when the staffers started relaying the report by phone to subscribers in real time. Soon hundreds of calls a day deluged the office, requiring five separate phone lines to handle all the information requests from paying customers across the city, then beyond.

When the new managers of the exchange figured out that there was a hunger for more data in the industry beyond just the current day's arrivals, they started collecting detailed information on all the ships that ever called at the port, then expanded the service to include a daily report that listed all the arrivals and departures expected for the next three days. When clients demanded even more, the report was expanded to look five days into the future, giving shippers, importers, exporters, and cargo owners unprecedented knowledge of their goods and supply chain. Each day the projections would be updated in case of changes, delays, or early arrivals. At the time, no other U.S. port had anything quite like this, and demand for this advanced planning tool went global. The Marine Exchange soon became financially self-sufficient. The little shack atop the warehouse was replaced with a proper watch tower that offered full amenities for the around-the-clock crew, with an expanded mission to serve the Port of Long Beach as well as Los Angeles.

The Marine Exchange remained essentially unchanged from this 1920s model until the 1990s, serving as a critical information gateway for goods leaving and entering the country. Steamship lines, harbor stevedores, trucking companies, railroads, and the ports themselves depended on that perch atop the warehouse, using its data and predictions to dispatch thousands of workers and vehicles every day, plotting the course of cans of tuna from Thailand, shoes from China, coffee from Colombia. All this information was stored in-house: decades of shipping data going back to 1923, with active shippers and vessels summarized on note cards coded with different colors representing various types of ships: tankers, bulk cargo, fishing craft, and passenger liners. Later, new categories and colors had to be added for container vessels and specialized car carriers. Radar and radio links to ships came online in the fifties and sixties, followed by identifying transponders similar to those used on aircraft. The cards representing a day's vessel traffic in Los Angeles and Long Beach were displayed on a rotating barrel-sized drum that allowed the marine information specialists to single out and update at a glance hundreds of daily shipping movements in the channels, anchorages, berths, and servicing docks. In an analog, pre-Internet age, the homemade spinning barrel and the decks of cards served as a sort of mechanical computer. (It was such an ingenious system that, when it finally gave way to a computer-based version, the Marine Exchange custom software began as a virtual representation of the old analog system, right down to the color coding of the cards.)

For all its ingenuity and value, however, the Marine Exchange suffered from one missing piece that affected capacity, safety, and order at the port. The exchange lacked the authority to use its wealth of time-sensitive information to act as a marine version of air traffic control, even though the increasing business of the port created a pressing need for such a service. The Southern California shipping lanes, filled with ever larger and more powerful ships, had by the late seventies and early eighties experienced escalating numbers of accidents and many more near misses—up to two hundred a month. As the trans-Asian cargo traffic multiplied and container ships began to increase in number and grow in size, the waters near the ports became a treacherous free-for-all zone. Incoming cargo vessels and tankers would spot other ships approaching and race to the harbor breakwater, directing their own courses, cutting across rival ships, trying to be first to slip inside the opening in the protective stone barrier. The port's first-come, first-served system then in place, and the overload that occurred when there were more ships than berths available, made conflicts and dangerous games of maritime chicken inevitable.

Coast Guard veteran Reid Crispino, who would join the staff at the Marine Exchange after leaving the service, recalls his reaction the first time he saw those old-school, free-for-all lines of ships jockeying to get in first. "This is crazy," he told a colleague. "But nothing's going to change until something terrible happens."

Something terrible did happen, though in Alaska, not Los Angeles. A tanker ship laden with Alaskan crude and bound for the Port of Long Beach struck a coastal reef in March 1989, rupturing the hull and spilling as many as 38 million gallons of oil into pristine fisheries and habitats. The running aground of the *Exxon Valdez* marked one of the worst environmental disasters ever in U.S. waters—and one of the most preventable, given that the outcropping the ship rammed was well-known to mariners and charted. One of the many lessons learned from the multiple human and system failures that contributed to this disaster was that a land-based, radar-driven vessel traffic service for directing ships safely near ports and coastal areas could have easily detected the ship's errant course and prevented the spill. The absence of such systems at all but a tiny handful of ports became a national scandal, and bringing them online became a national priority. By 1994 the Marine Exchange, newly charged with a vessel traffic control mission in partnership with the Coast Guard, moved to its current site atop Angels Gate, far from the signal clutter that made the old warehouse roof a poor spot for modern radar dishes.

Debbie Chavez joined the exchange during this transition time. Charged with helping with the switch from the old cards-on-a-barrel system to modern computers, she eventually became head of the entire marine information service division at the exchange. The modernization (and Chavez's career) paralleled the emergence of the Los Angeles–Long Beach port complex as the leading gateway for America's consumer economy—the place where the door-to-door dance is at its most complex, and where overload is now a daily reality, although in ways no one could have imagined back in 1994.

On February 13, 2015, the maritime information manager arrives at Angels Gate in the gray light of early morning. Fingers of wispy mist stretch out over the waters below the Marine Exchange, enough for her to feel the gloom and winter chill, but too thin to conceal the long line of waiting ships. The cargo vessels stacked high with Lego-colored containers stretch down the coast as far as she can see, a twenty-mile-long queue of goods and food and fuel and delay. The line will only grow longer this day, she knows. But the goods-movement ballet has to go on, and it starts this day, as usual, with Debbie Chavez, her five marine information specialists, and their phones. It's time to crank up the Boiler Room.

That's what they call it during those busy morning hours after dawn in Chavez's warren of offices staffed by men and women with phones glued to their ears. Tucked out of sight from the breathtaking panoramic views that dominate the traffic control room, the information service's resemblance to a telemarketing sweatshop or election phone-bank boiler room at these times is obvious. And boiling it is, as port congestion has never been worse than this month's backlog. It had been bad at many ports around the country lately, but murderous on the West Coast, where the longshoremen's union has been working without a contract for nearly a year, unable to come to terms with the shipping lines. The shippers, truckers, and union workers pointed fingers at one another for causing the ocean traffic jam. Although the Marine Exchange is neutral—the Switzerland of the seafaring world, as executive director Captain J. Kip Louttit, likes to say—the rancor had rendered Chavez's job all the harder. All her contacts were on edge.

She settles behind the computer monitors at her desk and tackles the first task: pinning down the expected arrivals, verifying that they are on schedule as predicted the day before. Chavez looks over the list and begins making her calls, starting with the nation's oldest and largest shipping agent, Norton Lilly International, which keeps an office in Long Beach, close to the conjoined ports and their clients. Norton represents owners of four of the five expected arrivals this day, including one of the biggest ships calling on the port complex, the *MSC Flavia*, a Mediterranean Shipping Company vessel capable of carrying 6,200 standard forty-foot containers.[2] Today it carries a mix of clothing, electronics, shoes, and a panoply of other goods from Europe and Asia bound for Target, Macy's, Kohl's, and thousands of other businesses near and far. Many steps remain in the ballet before any goods reach their ultimate recipients, the BCOs (beneficial cargo owners) in shipping parlance. The congestion delays have them all anxious, angry, and wondering if they should instead have shipped via some other port—or some other method, such as air freight.

The *Flavia* has just arrived in the vicinity and is working its way to an anchorage outside the Port of Los Angeles, where it will, like almost every other ship, have to wait for a berth. Chavez and her staff soon verify that four other container vessels have arrived early in the morning, two for Long Beach, one more for LA, all three assigned to "waiting-room" anchorages outside the ports proper. A fifth container vessel has arrived but its anchorage and port of call remain undetermined, as it carries freight for both LA and Long Beach, a complication of the new alliances among shippers that contributes to overload. Eight more ships are expected to arrive today, Chavez and her staff verify, but they are still at sea, hovering outside the two-hundred-mile limit and its pollution regulations, so they can continue to burn their cheapest, dirtiest fuel. This also keeps them far outside the twenty-five-mile coastal zone in which they must contact and accept direction from the Marine Exchange traffic controllers.

Next Chavez has to verify the day's "shifts." These are the ships moving between remote anchorages and berths inside the port terminals for loading and unloading. Only two vessels will be making it to berths today, neither of them container ships. Three others will be shifting to anchorages while they await spots at one of the thirteen massive container terminals at the two ports.

Chavez passes all this new information on to the traffic controllers stationed in the Watch, whose computer-rendered images of the ports begin to pulse with alerts about the pending moves. This system is a dramatic improvement compared to the analog days, when Chavez would walk over to a window dividing her operation from the Watch and write the updates on the glass with a grease pencil. She had to write backwards, so the controllers could read it from their side.

The updated arrival information next goes out to the port pilots so they can use it to plan their staffing, and to all the other

Marine Exchange subscribers up and down the supply chain. Then Chavez moves on to the rest of the day's and week's ship movements so that the advance reports can be updated and all the many players at the port, from terminals to longshoremen to pilots to crane operators to rail and truck haulers, can plan the days ahead, staffing the right number of people, preparing the right number of freight cars and trailer chassis, assemble the right number of gangs to unload the ships, to fuel them, to supply the vessels with food, water, and equipment. So much depends on Chavez's reports, but congestion at the ports is taking its toll, the worst delays and backlog she's seen in twenty-one years. That's how long it's been since this working mom born, raised, and still living in her hometown of Carson, California, left her bank teller job to dive into the world of port traffic.

The shipping snapshot she and her team crafted for February 13 reveals the depth of the overload:

- Pending arrivals: a total of thirteen ships are expected to arrive on Friday the thirteenth, with five more expected on Saturday, ten on Sunday, and thirteen on Monday. Of those vessels, twenty-three are container ships. The rest are tankers, bulk cargo vessels, and passenger cruise ships. Just over a week out, a total of 108 ships are expected to arrive.
- Pending shifts: fifteen vessels are supposed to move between anchorage and berth from Friday through Monday.
- Pending departures: thirty-seven vessels are supposed to leave the twin ports between Friday and Monday, fourteen of them container vessels.
- Total number of vessels at port: ninety-two ships overall are present, fifty-nine of them berthed at terminals. Of those waiting to get into a berth for unloading, loading,

or both, twenty-seven of them are backed up due to port congestion (the other six are awaiting supplies, fuel, or repairs).

One of the twenty-seven ships stalled by overload, the 4,000-ton container ship *MSC Charleston* has been waiting three weeks at anchor. Another ten of the ships delayed by congestion have been waiting ten days or more. The largest ship in either port, the *COSCO Denmark*, carrying 7,000 containers, began day five of its wait for a berth in Long Beach this day. In all, 19 container ships with a combined capacity of more than 80,000 containers are stalled in congested anchorages along with six break-bulk vessels and two general cargo ships carrying such items as fabricated steel girders, earthmovers, bulldozers, and aircraft components too large for containers. Those stalled ships hold billions of dollars' worth of cargo, all of it either late for delivery or soon to be overdue. Businesses with short time frames—candy, toy, and the gift businesses awaiting goods for Valentine's Day—are losing millions. Sale items already advertised are not on department store shelves. Now the springtime–Easter holiday products that should have already arrived are missing in action, too.

A year earlier, cargo ships faced little or no wait for a berth, and the turnaround—the time it takes to reach a terminal, unload, reload, and depart—was a point of pride for both ports as well as a source of competitive advantage. At that time, ships averaged less than three days in port, with small ships moving out sooner than that, and larger ones taking longer. Now Friday the thirteenth has doubled turnaround times for some shipping lines, and tripled or quadrupled those times for others.

Part of this arose from the long-standing labor dispute at the port and bickering over contract terms that had led to alleged slowdowns by workers and lockouts by shippers. Everyone from the White House down to the mayors of the two port cities have expressed growing frustration at the seemingly endless feud with such potentially dire consequences for the economy. Yet the high-stakes contract argument did not erupt over such straightforward points as money or automation. Instead the battle began over the more arcane question of whether the arbitrator who settles disputes between labor and management should be appointed for life or have a fixed term. (The shipping lines, who like the current arbitrator, want him to stay in place forever; the union, not so much.) But the long-term problem of congestion has more behind it than the lack of a longshoreman's contract. Bigger ships hauling more cargo have been taxing port operations and the trucking industry for years with the sheer bulk of shipments flooding the docks when big ships arrive in clusters. And the new trend of shipping alliances has made handling the larger loads on each ship more complex, as it has required some ships to stop at multiple terminals.

Another major factor has been the trucking foul-ups caused when shipping lines unilaterally decided to stop a decades-old practice of supplying the wheeled chassis that turn containers into semitrailers. The decision saved the shipping lines money but created havoc as the ports and freight-hauling companies struggled to set up alternatives—no easy matter when the majority of the trucks servicing the port are small owner-operated businesses working on contract. For much of 2014 and 2015, chassis shortages drove the port truckers away, imposing onerous delays on those who remained as hundreds of thousands of containers piled up in dockside mountains, undelivered. Truckers who once simply arrived at the port, picked up a container and chassis combo, then left suddenly and had to make multiple stops, first to find a chassis, then to pick up a container somewhere else at the port, and then wend their way back to the exit. Predictably, this created in-port traffic jams, unproductive down time, and a chain reaction of delayed cargo. This sort of congestion and overload at ports is a nationwide phenomenon, not just a Los Angeles thing. New York/New Jersey, Oakland, Virginia, and others have been hit hard, too, as well as the intermodal rail terminals in the Midwest that handle the goods moving to and from the coasts. Trucks sitting and waiting to pick up and deliver containers contribute to excess air pollution and carbon emissions, and impose huge costs on the consumer economy: $348 million wasted a year, a figure that includes nine million gallons in wasted diesel fuel and fifteen million hours of wasted worker time.[3]

Settling the labor dispute with a new contract would provide short-term relief, but would not address any of these other overload threats.

"Our position is that we're the honest broker. We take no sides on any of this," Chavez says, looking over the long list of ships parked, knowing that every extra day in port costs money and jobs up and down the chain. "But it's not taking sides to state the obvious: this is a mess. Everybody loses."

Up in the Watch, gesturing at the long line of ships snaking south down the coast, general manager Reid Crispino says, "See for yourself. It goes all the way down to Huntington Beach."

Oustide the picture window, the twenty-mile-long line of waiting ships stretches as far as the eye—or the telescope—can see. It vanishes into the distance.

Crispino falls quiet a moment as a colleague speaks softly into a radio microphone, confirming a container ship's position in the distant contingency anchorages. On the large multicolor computer display, which combines radar, GPS, and transponder signals from the ships with a map overlay of the ports, the dot representing the ship on the radio seems to be floating inside a force-field bubble. The screen image represents the circular buffer zone that each cargo ship in the anchorages is supposed to maintain to avoid collisions. That bubble is 1,200 yards—more than a half mile—in diameter. No other ship may enter that circle. However, it's the Marine Exchange's vigilance, not force fields, that keeps ships separated in congested waters. This is a necessary protection given that these massive vessels are surprisingly fast but not very agile. Collisions are almost unheard-of and near misses are a fraction of what they were in the eighties, Crispino says. "If any good came out of the *Exxon Valdez*, you're seeing it on that screen."

Crispino, like so many others working at the port, has been a lifelong mariner. He grew up in Anaheim, California, a few miles from Disneyland, but his landlocked home only heightened his attraction to the sea. He learned to sail before he could drive, navigated solo to Catalina Island thirty miles offshore when he was thirteen years old, and joined the Coast Guard at age seventeen. In the service, he worked fishery patrols in Alaska, was engaged in the search-and-rescue branch in California for eight years, returned to Alaska to help with the *Valdez* oil spill cleanup, and retired after more than twenty years of service. He joined the Marine Exchange in 1997, happy for a more stable home and work life at this center of the shipping universe and consumer economy. Still, he admits that, like almost all the veteran mariners here, he still misses the exotic travel, the feel of a deck underfoot, the spray in his face. Then he thinks how miserable those crews must be trapped out in those anchorages, and working with a computer display of ships rather than the real thing doesn't seem so bad after all.

The radio squawks again. It has been a slow day—congestion means less to do on the Watch, not more—but it is time for a shift. Berth and course information are exchanged and confirmed by radio, and the traffic controller calls the pilot's office over at Long Beach.

"At least someone's going to unload today," Crispino says as the ship in its bubble begins to move across his screen.

Chapter 9

THE BALLET IN MOTION

The little pilot boat bounces and slaps the waves as it sidles next to the towering cargo vessel as they match speed and course, the little boat looking hopelessly fragile next to its eight-story, windowless counterpart—an egg challenging China's Great Wall.

The cargo vessel *Wisdom Ace* looks more like a floating building than a ship, which makes sense: it is, in effect, a floating parking garage, a very specialized sort of cargo ship known as a RO-RO (for "roll-on, roll-off"). It's a car carrier, and the beauty of this type of vessel is that the cargo can be driven on and off rather than craned into place like containers. Today the *Wisdom Ace* carries 5,000 Mercedes cars and SUVs from Germany, parked and lashed bumper to bumper on the ship's many low-roofed decks. The Jacobsen Pilot Service boat is there to deliver a pilot, whose job is to bring the big ship and its costly cargo safely in to dock.

The Jacobsen boat steers close enough to count the rivets in the big ship's looming expanse of steel plating; close enough to lean out and *touch* the rivets.

A third of the way up the featureless expanse of the *Wisdom Ace*'s hull—which is long and wide enough to cover two football fields—a door opens out to nowhere about twenty feet above the roiling water. This is the pilot's door. A rope ladder is dangled from it, and two men peer down, one of them waving.

"That's my cue." Bob Blair stands up and stretches. He had been resting in the sun-drenched stern of the boat, maybe even dozing. It seems mariners can sleep through chop and wind and the approach of floating parking structures, no problem. He had been up since three that morning, after all, and this would be the third ship he piloted this day through the Port of Long Beach's narrow and quirky channels, which feature a bridge so old it must wear a diaper to catch falling bricks, and submerged obstructions with such ominous nicknames as "the Can Opener." Blair, like his fellow pilots, knows all these idiosyncrasies, dangers, twists, and tight spaces as no visiting helmsman could. It's his job to find the safest course to bring the cargo across the last few watery miles of a global crossing.

The sandy-haired man with the blue dress shirt, necktie, and immaculate black windbreaker steps up to the portside gunwale of his boat, adjusts the zipper on his jacket, and then leans out over the whitecaps. Without breaking stride, Blair grabs the rope ladder and steps up onto its nearest rung, a single fluid, casual motion, as if he were stepping from street to curb rather than leaping onto a steel-walled monster that could kill him in an instant if he slipped. He nimbly scales the ladder and disappears through the door as the pilot boat peels off, dodging the frothing wake kicked out from the big car carrier.

Less than forty-five minutes later, Blair eases the *Wisdom Ace* through the port and into its berth at Pier F so the stevedore gang can begin unleashing the cars, driving them down the ramp and off to a holding area. Luxury cars and SUVs stream off the ship like spectators leaving the ballpark after a game.

It is almost 11:00 a.m. by now, and Blair awaits word on his next assignment: piloting a container ship out of China to the backlog anchorage.

"Every port is different," Blair observes. "These ships call all over the world. There's no way their crews can know how to get in and out of each and every port. There would be slowdowns or accidents all the time. The ports require pilots because we know our ports. We cut the risk. I wouldn't say we're treated like royalty when we board a ship, but they're usually glad to see us."

Port pilots are among the elite of the shipping world, part of its hidden choreography, essential but unheralded professionals who keep the cargo moving through its tightest, most congested spots. They board the ships as they arrive and take them to their berths, then guide them back to the anchorages when they depart. Tugboats assist with the large ships, teams of them on windy days, as the immense cargo vessels act like sails because they present such a large surface for wind to act upon. A 15-knot breeze can generate one hundred tons of force on a container ship, enough to blow it sideways or cause its tail to swing around. Tugs have to exert that much pressure on the downwind side to counteract that force—it would take two or three to match that 15-knot breeze—all under the direction of the pilot. Pilots never touch a control but stand on the bridge and instruct the crew on speed, course, and turns. Blair says there is often a language barrier, but the simple commands are known by all in any language, and he usually makes his meaning clear. It is a formal and ritualistic process at times, the welcome sometimes lavish, with offers of food and beverages, and at other times it's more tense and wary. But the pilot's word is always final while in the port waters.

By way of comparison, port pilots are the surgeons who swoop into town to perform some complex, difficult procedure on a patient whose regular crew of caregivers does all the prep work and aftercare. Or perhaps the pilot's role is best likened to the short relief pitcher in baseball, brought in to record that last out or two with everything on the line. It's stressful but over quick, the end of an ocean journey that lasts weeks on the open water, but where the risk is rarely greater than when the ship is a handful of miles from port. The pilots are the closers of the door-to-door world, a breed apart, with coveted jobs that pay some of the highest salaries in the marine business. Certainly the compensation is comparable to surgeons and professional ballplayers, ranging from about $300,000 to nearly half a million dollars a year.

Most are career mariners. Blair came up as a tugboat pilot, those essential helpers when maneuvering big ships through narrow port channels. He eventually applied for a pilot's job at Jacobsen, which tests its applicants mercilessly and, given the rewards, can afford to be choosy. He's been there eleven years, which ranks him among the least senior of the pilots. Jacobsen is a destination job in the port piloting field, and it's run like a family business despite taking part in 7,000 ship moves a year: everything in Long Beach, plus the U.S. Navy ships moving in and out of the nearby Naval Weapons Station Seal Beach, plus the small number of American-flagged vessels calling at the Port of Los Angeles, mostly oil tankers.

Jacobsen is a unique presence in the port pilot business, a private company where the seventeen pilots have ownership interest, compared to most other operations, where the pilots work in a union shop or are municipal employees, as in the Port of Los Angeles. Jacobsen has held the piloting contract in Long Beach since 1922, when the company was founded by an immigrant Norwegian fisherman, and it has been run by three successive generations of Jacobsens ever since. Blair's boss, Captain John Strong, head pilot and vice president at Jacobsen, has been a pilot for thirty-two years and a mariner even longer, serving on research vessels out of San Diego (where he worked for the man who discovered the wreck of the *Titanic*), supply ships out of Tahiti and the Virgin Islands, and oil tankers in the Gulf of

Mexico. He has always loved to be at sea—a compulsion, he says, fueled by a restlessness that inevitably seized him if he stayed land-bound too many months. Through his wanderings he found three wives and lost two of them, felt the pain of coming home to kids who did not recognize him, and then found that the pilot's life could be the best of both worlds. A pilot can be on the water all day and yet home for supper every night, which meant his third marriage has stuck now for twenty-two years and counting.

"I'm the kind of guy who has to be at sea and who has to be married," he says. "So this has worked out well for me."

Of his job, he says it often boils down to this: "We are the only ones who can tell a ship's captain: No."

The pilot's job has grown more complicated during his career, first due to the containerization revolution, and most recently because of the trend toward ever-larger mega-ships. Now the ships are so big that when the pilots sail up the channel to the Long Beach inner harbor, they cannot even see the water or how close the ship is to the banks. The clearance in the tight spots could be as little as fifty feet on either side of the bigger ships, which, by comparison, would be like driving a car for miles with less than two feet of clearance on either side. (Drivers are accustomed to much more clearance than that: typical freeway lanes are twelve feet wide, twice the width of the average car, so that on a two-lane highway there's usually about six feet of separation side to side between vehicles.) With clearance proportional to the tight inner channels of the Port of Long Beach, a twitch of the wheel would crash the car—or the container ship—into a wall.

Technology has come to the aid of the pilots, while the job remains a curious mixture of primitive and advanced. The rope ladders are right out of the *Mutiny on the Bounty* days, but Bob

Blair also packs a state-of-the-art wireless GPS tablet with live feeds of port conditions, traffic, tides, winds, and depth in order to navigate those tight blind spots without incident.

The inner harbor presents particularly tight quarters with very busy port terminals inside. Ships headed there pass beneath the aged Gerald Desmond Bridge, which links the harbor to the critical-goods corridor of the 710 Freeway. This is the bridge with the diaper: nets underneath have hung there for years to catch crumbling concrete that has been raining down from the bridge as the structure slowly disintegrates from heavy use. The port officials estimate that 15 percent of U.S. consumer goods cross that bridge every day, although it is too fragile and too low at 155 feet for modern cargo ships. Radar and radio antennae have to be retracted on the tallest vessels, navigation masts folded, flags lowered. The quarters are so tight that, during the passage of one monster cargo ship stacked high with containers on deck, Strong actually had to stop the vessel and clamber atop the containers with a yardstick in hand to make sure the ship could pass beneath the bridge safely. It did, with inches to spare—at low tide.

A $1.5 billion replacement bridge with more lanes, bike and pedestrian ways, and a 205-foot clearance to accommodate larger ships was expected to open by 2018—over budget and two years late.

The inner harbor area has another bottleneck near a power plant, where a water intake valve extends into the channel. This is the spot the pilots call "the Can Opener" because it could puncture a ship's hull and rip it open like a tin can if not avoided. It's slated to be removed eventually to create more leeway for the big ships.

"Every trip is different," Strong tells his pilots. "Different ship. Different crew. The weather changes. You can have the best voyage plan in the world, but you have to go in with no assumptions. Because I guarantee things will change. And you'll have to respond."

When the pilot's job is done, the cargo still lingers at the port—too long for anyone's liking during times of congestion and overload—but it stops being ocean freight at that point and becomes land freight. And it is here that the ingenious design of the modern container ship—as opposed to the age-old design of bulk cruisers with traditional holds, ramps, and hoists—comes into play.

The old way—dating back three thousand years to the ancient Phoenician traders with their monopoly on precious royal purple dye, and continuing right into the post–World War II era—was to unload a ship as a moving company would empty a house. Stevedores would manually carry out furniture, boxes, and sacks, loading them on—depending on the era and place—trucks or boxcars, horse carts or camel caravans. The modern way is like a factory assembly line, all huge machines and rapid repetition, enabled by the complementary symmetry of container ship and shipping container. They're all standardized. And so every move to load or unload is the same. This makes for a mechanized dockside world that is fast and fearful and dangerous—and, aside from contract disputes and other hitches and obstacles, incredibly efficient. Humans need not enter the cargo hold of a container ship. Indeed, it can be perilous for them to do so, and while containers are being unloaded, only a slip and a fall will put a human body in the mix, never with good results. In 2014, a dockworker tumbled through a gap in containers aboard a small Panamanian freighter at the port, seriously injuring his legs and back. There have been deaths at world ports from similar mishaps.

The ship's design enables this hands-free efficiency. A modern container ship is built in sections called cells, which is why such a vessel is also called a cellular ship—not because of its wireless capabilities, but because it is assembled like a cross section in reverse. Imagine taking a passenger car and, beginning with the front end, slicing it from hood ornament to ground level, then carving one after another vertical section, each about a foot wide. On the gargantuan scale of container ships, these are the sorts of sectional building blocks that are crafted separately, complete with all the internals—conduits and corridors, plumbing and wiring, support beams and hatches. At the shipyard, these cells are then placed together, their components connected and the metal structures bolted and welded into whole ships. And the secret sauce inside the vast hollow interiors of these ships, built into every cell before they are assembled, is the system of container guide channels: vertical rails that become part of a ship's superstructure and sized exactly for standard forty-foot shipping containers. The rails guide the big metal boxes into place in orderly, tight-packed, and secure rows right down to the deepest level of a vessel, so cans don't have to be painstakingly positioned. They are just lowered, slid, and locked into place.

This simple, ingenious design allows not only for rapid-fire loading and unloading but also meticulous cargo planning. Each container's place on a ship is identified by three map coordinates: the vertical row, measured from number one at the bow of the ship, increasing toward the aft; the horizontal tier, with number one starting at the bottom layer of containers and moving up in number toward the top of the stack; and the slot, which measures container position from side to side, with even numbers on the port side and odd numbers starboard. Even with 6,000 containers on board, the position of every piece of containerized cargo is known (although the mega-shipping alliances have undermined the efficiency of this system by jumbling together cargo from different shipping lines).

Once unloaded, each of the 22 million intermodal containers in the world can be identified by their internationally standardized reporting mark, and by embedded radio-frequency identification (RFID) transmitters. The identity of each container is supposed to be verified multiple times by the stevedore gangs so nothing gets lost in the vast and confusing topography of millions of stacked containers in yards and terminals at the port.

None of this would work if the number one design principle of the modern container ship was not ease of unloading. Fuel efficiency, power, safety, comfort, emissions—all are important concerns to the shipping lines—but if the containers can't flow off a ship with ease and speed, the vessel is worse than worthless. It becomes a money pit.

America does not build such ships: Japan and South Korea lead the world on the creation of ever larger container ships. Daewoo Shipbuilding & Marine Engineering in Korea, one of the top five shipbuilders in the world, has 46,000 workers building 100 ships and oil platforms at any one time, including one of the world's largest cargo ships, the Maersk Triple E. (The three *E*'s represent economy of scale, energy efficiency, and environmental performance.) This model of ship is 1,312 feet long and 194 feet wide, with the capacity to carry 9,000 cargo containers per trip. So far they serve only Asia and Europe, as they are much too big for even the newly expanded Panama Canal locks or any North American port. Some of the berths in Los Angeles and Long Beach could dock them, but not unload them: they lack cranes with enough reach.

These mega-ships require only fifteen crew members to sail across the world, and cost $190 million apiece. Maersk ordered twenty. Daewoo has the capacity to build twelve simultaneously.

Once a container ship makes it out of the waiting room anchorages and reaches a container terminal, the unloading becomes another exercise in multi-ton surgery. Mammoth cranes capable of spanning the 170-foot-wide ships are positioned up and down the length of a vessel to begin the extraction of the containers. There are 140 electrically powered ship-to-shore cranes at the twin ports, a distinctive sight on the skyline, particularly when they're idle and the boom arms are pointed skyward, like soldiers firing a twenty-one-gun salute. The bright red and blue crane towers run three hundred feet high and will soon be taller. The ports are painstakingly raising them sixty feet by giving them longer legs to accommodate larger, taller container ships, at a cost of a million dollars apiece (versus $10 million for each new crane). Almost all are imported from China; America makes neither the ships nor the equipment for unloading them, and they have to be transported already assembled on specialized cargo ships.

The crane operators say their job is among the best at the port, if also one of the most stressful. It's also one of the most financially rewarding, as the most skilled crane operators can earn nearly as much as the port pilots. "It's technical, it's exciting, it's rewarding, and it's important," a hard-hatted crane jockey shouted over the din of her machine as she stepped into the elevator that lifted her three hundred feet up to a catwalk leading to the operator's cabin. "I love it."

The crane cabin rides on a trolley that lets it slide back and forth on the high horizontal boom that stretches over the width of a container ship. Metal cables spool up and down beneath the crane operator's feet, from which the "spreader" is suspended. The spreader takes the place of the old-school hooks that dangled from the cranes of the past. It attaches to the fittings atop the containers, with expanding metal grips that lock firmly into place and then retract after a "can" is lowered to the pier surface. The operator directs the cabin out over the ship while sitting in a moving armchair surprisingly reminiscent of Captain Kirk's command position on the original *Star Trek*, buttons and switches beside each hand. The operator grips in each hand a joystick used to lower and raise the spreader until it's lined up properly with the container top. Wind and the ship's roll in the water add to the challenge, and the complexity of the task has so far kept humans in the loop rather than robots, although a couple of terminals are beginning to use automated cranes and vehicles to move containers after they come off the ships. Most of the time the operator is staring downward between his or her feet through a window in the cabin floor, watching intently as the spreader drops into position.

Red and amber lights give way to green when the spreader is properly gripping a container, and the operator quickly lifts the can. There's nothing slow or deliberate at this point: the move is a sprint, not a jog, even though the object being lifted could easily weigh ten tons or more. The cabin rides back, away from the ship and into position over an empty trailer chassis, where the operator lowers and drops the can. As soon as the container is settled, the crane detaches and immediately zips back for the next container while the newly laden truck leaves and the next empty chassis is driven into position.

Crane operators at the California ports can average between twenty-five and twenty-eight containers an hour—just over two a minute. The highest paid and most sought-after operators routinely handle more than thirty cans an hour and can earn $250,000 a year with a thirty-hour workweek. They move more cargo in two minutes than the old bulk cargo stevedores could unload in an hour. And yet, even with four cranes working the bigger ships at once, and all operating at peak speeds, a 6,000-container delivery takes 54 hours to unload entirely, not counting time to reload (even when many of the outgoing containers from American ports tend to be empties).

During the labor-management dispute at the ports in late 2014 and early 2015, the crane work slowed dramatically, contributing to congestion by lengthening the time it took to unload each ship. When the contract was finally settled, it took nearly three months to work through the backlog. And yet the problem of congestion remained, for the bigger ships have a longer-lasting effect on system overload than labor unrest: more containers on each ship mean more hours to unload.

When the crane operator's work is done, the terminal gangs of longshoremen take over, moving the cans into temporary holding areas, where towers and pyramids of the different-colored containers amass until the proper truck or train is ready to be loaded. Marine clerks sort through the mazes of containers, some of which are difficult to find because of malfunctioning RFID devices or containers placed or logged incorrectly. The containers are moved in and out of the mountainous stacks by rubber-tired gantry cranes—smaller versions of the ship-to-shore cranes—which are mounted on inverted U-shaped frames riding on giant tractor tires instead of towers. As part of the LA–Long Beach green ports initiative, several of the terminals are replacing these traditionally diesel-powered cranes with electrical models. Most of the port terminals in the U.S. have human drivers on their gantry cranes, but driverless automation is being developed to take over this task and cut labor costs. In the twin ports of Southern California, two shipping terminals are rolling out autonomous gantry cranes, so far with mixed success.

The terminals, many of which are subsidiaries of the shipping lines, are charged with moving those containers out of the ports as quickly as possible, but once again overload has complicated the job. Just under a third of the containers depart via dockside rail (or near dockside, after a short truck ride). The Alameda Corridor could handle twice the number of containers currently moving through it, but lack of rail capacity inside the ports represents a bottleneck limiting the number of trains moving cargo through the corridor. Plans to expand the capacity with construction of a new rail yard near the port have been stymied for years. This project, dubbed the Southern California International Gateway, faces neighborhood opposition, environmental complaints, and a lawsuit filed by the City of Long Beach against the City of Los Angeles, which gave the Burlington Northern & Santa Fe Railway permission to build the facility. BNSF, the nation's second-largest freight railroad next to Union Pacific, was purchased by billionaire investor Warren Buffett's Berkshire Hathaway Inc., in 2010 for $44 billion.

Given the limits on rail movement from the twin ports, the next stage in moving our stuff door to door is all about trucks. About 70 percent of the cargo moves out via drayage trucks, the short-haul semitrailers that jam the ports and surrounding roads, each one carrying a single container. These trucks are a major source of air pollution and traffic congestion in the region.

There are about 10,000 full-time and 4,000 part-time drayage drivers working out of the Long Beach and Los Angeles ports, and each day they swarm the marine terminals. It's difficult and not always rewarding work, as picking up containers at the ports is a daily exercise in patience and dockside traffic jams even on the best of days. Drayage drivers for the most part are paid by the load, not by the hour, so idle time is a loss for them. The drayage truckers are an important link in the national goods movement system, never straying far but performing the essential service of bringing the still-containerized goods to nearby rail yards and transmodal train terminals,[1] product distribution centers, warehouses, and long-haul trucking operations. Except for a few large companies with their own trucking fleets—Walmart, the big food and beverage companies—the next move after drayage for most of the goods that come to America through ports—and from American manufacturers as well—is handled by for-hire trucking fleets and logistics companies. These are traditionally known as the "common carriers."

So the next stop for most goods out of the Southern California ports are close-in distribution facilities. The nearest network of these is in Carson, the little town on the south end of Los Angeles that Debbie Chavez calls home, where developers are dying to build the next new National Football League venue (commuters on the nearby, already congested 405 Freeway fervently hope they fail), and where a significant portion of the city lives and breathes and profits from its key role in the door-to-door system.

The Watson Land Company is a major force here. It sells that most valuable of quantities in a dense urban area: space. Or, rather, it leases that space on land far too pricey to sell because of its strategic location near the port, rail lines, freeways, and markets. Every port and city has such places, but Watson is a special case. Its owners are descended from the Dominguez Family, which was granted a huge swath of what is now Southern California by the king of Spain. Rancho San Pedro once stretched from where the ports are now all the way north to the edge of downtown, a vast empire of what was then undeveloped coastal vistas and rich farmland, though it has been broken up and sold over time. But the descendants still own 1,500 prime and priceless acres where a network of trucking companies and cargo receivers now reside in industrial and commercial buildings built and leased by Watson. Macy's, Maersk, International Paper, Alcoa, Mercedes, Bristol Farms, Herbal Life, Dermalogica, U.S. Customs, and a host of logistics companies few would recognize bring their cargo in and through this sprawling campus. Pilar Hoyos, an executive at Watson and another member of the LOLs, sees her company as part of the hidden glue that holds the transportation system together, one of the layers of goods movement as essential as it is little known. She loves to visit her tenants' operations whenever she can, she says, because seeing the flow of goods, the constant change in taste and trend, is like glimpsing the inner workings of a vast and mysterious machine. "And all can be traced back to the old ranchos and the king of Spain and a family's decision to hold on to this piece of it."

But space is only part of the puzzle. If most companies don't have their own trucking assets, how do they move the goods to their department stores, their shops, their factories? In years past, businesses would make their own arrangements, hire truckers, or haggle with railroads. Some still do. But the trend now is to farm that work out. Companies such as Frontline Freight in the nearby City of Industry work for Watson's tenants and other businesses across the nation; they are one of a new and growing breed of truckless, trackless transportation companies known as third-party logistics providers or freight forwarders. Frontline has a big warehouse and a computer bank staffed by a roomful of transportation data specialists. Each day the warehouse fills up with every sort of product imaginable, most of it from the ports: Chinese-made life vests, fitness equipment, pottery, barbecues, eight-hundred-pound aircraft parts, clothing, shoes, gourmet restaurant stoves, children's toys, Halloween costumes. "Here, look at this load," sales manager Ben Fauver says. "Tree stumps." A load of tree stumps had come in for use in some sort of outdoor display or movie set. "We get everything. And at the end of the day, everything's gone. We sweep the deck, and by the next morning it's full again."

What Frontline does—like hundreds of other companies in this growing "3PL" line of business—is arrange to receive the goods for an importer or other freight recipient (the goods can be domestic or imported, anything from anywhere is fine) and arrange to have the freight shipped to its final destination. That could be across town, the state, the country, or the world. Think of this business model as a travel agent or, perhaps closer to the mark, an HMO for shipping. Frontline uses the power of pooled customers and insider contacts to negotiate lower rates and faster shipping in exchange for a percentage of the fee. This is, increasingly, how the goods-movement system works, relying on brokers to put the pieces together—brokers not of goods but of movement.

The freight forwarders engineer a triptych for cargo—from port to drayage to carrier—with stops at close-in waypoints such as Watson. Next it's on to more distant destinations in the California desert, where hundreds of square miles have been transformed into a landscape of sprawling distribution centers (think everything from Amazon to Zappos and every company in between). Next rail, air, and long-haul truckers move the goods to the rest of the nation—on to our stores, our businesses, our hospitals and schools, and through the last mile to us. To our doors.

And it is here, where our stuff meets the road, that the most visible and constant overload kicks in.

Chapter 10

THE LAST MILE

In the cavernous basement of the Olympic Building, a line of boxy, dark brown delivery trucks rolls out to the early morning streets of downtown Los Angeles—a chorus of tires squeaking across smooth concrete. Five floors up in the president's office, Noel Massie allows himself a brief moment of contentment as he feels the building vibrate around him and then fall still with the last of his fleet's departures. This is the reassuring physical signal that his part of the never-ceasing, get-it-now economy has successfully turned one more notch on its endless loop—a cycle repeated this day at 2,000 similar United Parcel Service delivery hubs around the country and the world.

In the next eight hours this cycle will land 15.3 million packages on America's doorsteps,[1] where people will find what they need, what they want, and what they bought without ever leaving their homes or businesses. But Massie has no time to linger over the everyday marvel of delivering so much so fast, as he must push on to the next cycle: his people are already planning for the incoming packages that will soon be barreling back to the Olympic Building and all its many sister locations, ready for unloading, sorting, redirecting, and delivery—the stuff of tomorrow's doorsteps.

It is fair to say that Noel Massie's days are dominated by two things: trucks and minutes. He has too many of one, too little of the other, with 2 million shipments under his purview moving one way or the other hanging in the balance—every day. He is the door-to-door economy incarnate, although his official title is district president of United Parcel Service.

There used to be fifty UPS presidents in the nation, but a wracking consolidation in 2012 knocked the number down to a more efficient, less costly twenty. But these men and women had been the princes of America's leading door-to-door company. Imagine the White House eliminating 60 percent of its cabinet along with all its executive staffs: that was the level of transformation that shuddered through the company. But when the dust settled, Massie ended up with one of the biggest and busiest slices of the UPS pie in the world: the southern half of the state of California, from the Mexican border to the City of Fresno (plus Hawaii, southern Nevada and western Arizona). His headquarters are in the nondescript gray and brick building at the somewhat shabby corner of Olympic Boulevard and Sunbury Avenue in downtown LA, only one of many operating centers in Massie's purview, which includes an array of far-flung distribution centers, truck terminals, an international airport, and 20,000 employees. Now he is overlord to an immensely desirable customer base of Amazon-ordering, iPhone-buying, one-day-delivery shop-a-holics, along with many of the businesses that serve and sell to them. But his delivery nirvana is balanced against a landscape of traffic and sprawl seemingly designed to make his job of daily drop-off and pickup all but impossible.

"I am in the business of minutes," Massie says. "It's all about the minutes. If the plane leaves at seven, you either get there or somebody doesn't get what they need in time. Brain scans for someone's surgery. Tissue samples for the lab. You can't mess that up. Minutes matter in this business."

Before packages, before sorting and bagging and loading, before driving and delivering, there is the clock, a UPS president's true boss, Massie says. "Minutes make us or break us" is one of his mantras.

For all that, the man has an impish quality about him. He is focused but funny, balding but unlined at fifty-seven years, quick with a smile but also—he can't really help himself—a reflexive clock watcher, always checking the time. Massie's schedule begins each weekday with his 5:30 a.m. rise from bed at his family home in Yorba Linda, an Orange County suburb thirty-five miles south of his office. It ends with his departure from the Olympic Building twelve hours or so later. "I don't have a specific quitting time; I work till I'm done."

What does that work look like? On an average day, Massie's Southern California employees will make 1.2 to 1.3 million deliveries in Southern California, more than 8 percent of the UPS worldwide total, generating more than 8 percent of the company's total annual revenue of $58.2 billion.[2] He does this with about 5 percent of the UPS workforce (which is 435,000 worldwide, moving 6 percent of the nation's GDP).

A secret weapon makes this feat possible: a staff of 150 industrial engineers. This is the title UPS gives to the men and women whose job is to design the optimum route and order of stops that will get delivery drivers where they need to be when they need to be there while using as few minutes and miles as possible. The brown trucks are the symbol and the familiar face of UPS as far as the outside world is concerned, but the heart of the operation, the force that keeps the whole complex clockwork moving, is the army of engineers mapping and calculating morning and night. With more than 10,000 drivers in Southern California averaging 120 stops a day, in the most traffic-ridden, constantly changing urban sprawl in the U.S., Massie's troops face one of the toughest choreographing challenges in the door-to-door universe.

The first tool in the UPS engineers' arsenal is the built-in "telematics" data devices every truck and driver carries. This hardware relays each truck's performance information in real time to the engineers, who compare it to previous days on the same routes. With this data they can identify streets, turns, and intersections that are causing delays because of shifting traffic patterns, detours, or construction—even small delays drivers may not notice. The data lets them build more efficient routes for the next day.

Then there is the company's famous no-left-turn policy, put in place in 2004, when the engineers realized that drivers waiting to turn left with engines idling were burning significant amounts of minutes and fuel. By assigning routes that avoid lefts for 90 percent of a delivery van's turns, the company found it shaved 98 million minutes a year of idling time from its routes, which not only sped deliveries but also saved the company about 1.3 million gallons of fuel a year. Avoiding the left is also a proven safety measure, as traffic data shows that left turns are involved in ten times as many crashes and three times as many pedestrian deaths as right turns.

The industrial engineers' newest and most sophisticated tool is a computer program called ORION (a catchy acronym for a decidedly uncatchy 1,000 pages of computer algorithm known as On-Road Integrated Optimization and Navigation). No human can consider all the possible routes with brainpower alone—the variations for one truck with 120 stops in different locations with varying drop-off and pickup times yield a number too high to have a name (trillions just won't cut it). Rounded off, it is best expressed in scientific notation: 6.7×10^{143}; if you wrote this value down in normal notation, the number of possible routes would look like this: 6,689,502,913,449,135,000,000,000,000,000,000, 000,000,000,000,000,000,000,000,000,000,000,000,000,000,

000,000,000,000,000,000,000,000,000,000,000,000,000,000, 000,000,000,000,000,000,000,000,000,000,000,000,000,000, 000,000,000,000,000,000,000,000,000,000,000,000,000.[3]

Common sense and driver experience have been the routing tools used for most of UPS's century of existence, but a method of mapping with certainty the most efficient and effective route has been elusive. It's a classic mathematical conundrum known as the Traveling Salesman Problem. Now ORION can crunch that big number down to a short list of optimal routes that saves both minutes and miles, mapping out turns and tweaks that are too numerous for any human driver or engineer to compare unaided. The humans take that list, modify the routes that are supremely efficient on paper but make no sense in the real world, and the trucks are ready to roll. Massie says the results have been impressive, in part because the small savings each ORION route finds can add up fast when you've got nearly 100,000 delivery vehicles plying the world. Shaving just one mile off every truck's route can save the company $50 million in annual fuel costs; UPS expects up to $400 million in savings when ORION goes company-wide in 2017.

As complex as all this sounds, Massie's job would be much easier if that was all there was to it: just a simple problem of engineering. That's what he was studying thirty-four years earlier, a semester away from his degree and already interning at IBM. Then UPS plucked him out of his part-time job loading trucks and offered the industrious young man out of East Oakland a full-time gig with room for advancement. He never really stopped approaching his work as an engineer would, and the true daily task is much more than delivery and pickup. Those are just the bookends in the process, the publicly visible beginning and end points of a much bigger race. The company may be delivering 18 million parcels a day, but only 2.7 million are overnight air shipments. This means that, at any one time, the company is juggling 100 million or so packages (more during holidays) while they are in transit.

Routing all that requires a twenty-four-hour operation. In Massie's district—as in any UPS district—the cycle begins around 1:00 a.m., when the fifty-three-foot big rigs—"feeder trucks," in UPS-speak—move between cities and regions laden with ground shipments. Because UPS uses a hub and spokes system for both air and ground deliveries, few trucks haul parcels beyond a five-hundred-mile radius. A feeder truck bound for Salt Lake City from Los Angeles might stop at Las Vegas and meet a truck coming in from Utah. The two drivers will unhook and swap their trailers, then turn around and go home. Longer-distance shipments out of Southern California—about 80 percent of packages and documents—arrive and leave by rail, with the faster (and pricier) air shipments headed to the company's regional air hub at Ontario, California, the unlikely desert location that UPS has made into one of the dozen busiest cargo airports in the country.

The feeder trucks, trains, and planes meet up, crisscross the country, and bring the packages toward their destinations, ultimately landing at sorting centers and delivery facilities like Massie's Olympic Building. They are, literally, feeding the beast.

At 4:00 a.m., the night loading of the delivery trucks begins, preparation for the final stage in the package shipping process. Parcels that arrived earlier by air or feeder truck or were picked up by the delivery vans themselves are sorted, scanned, and incorporated into ORION's route-planning calculations, which are continually updated as new pickups arrive. While the sorted packages are being put on delivery trucks, the routes are finalized and downloaded into the drivers' tablets (UPS had deployed this tech years before the iPad came along). Then the iconic brown box trucks depart to complete their deliveries—the endpoint the customer at the doorstep actually sees. Finally the same drivers complete their pickups—three quarters of a million package pickups in Massie's Southern California district—and return to the network of operating centers, usually between 6:00 and 7:00 p.m. There, incoming packages are sorted by destination and shipping method and sent out by feeder truck, rail, and air to the proper UPS hub, and the process begins anew, sometimes with bare minutes to spare before a plane, train, or truck departure.

UPS has a panoply of businesses: it's a massive ocean shipper, although it owns not a single ship. It runs a separate freight-hauling trucking line, acting as a common carrier for businesses nationwide. It runs a drayage division that transfers goods out of the nation's ports, it's a logistics and freight forwarder service, it's an e-commerce consultant (online shopping being their main growth area as shippers), and it is classified as an airline, with enough planes to put it in the world's top ten carriers. UPS even operates its own bank. But the main business, and the main source of revenue, remain what they have been for more than a hundred years: UPS solves the last-mile problem with its doorstop deliveries.

When it comes to transporting humans, there is no similar solution that comes right to our doors. Trains, buses, subways, and trolleys can move many people at once (as UPS vans can move many packages at once), but these more efficient human conveyances have not been able to beat the car—automotive inefficiency, cost, and death toll be damned. The car alternatives cannot affordably take a person door to door. You have to get to the station or the bus stop or the platform. That metaphorical last mile remains a barrier of time, distance, and inconvenience between transit and the traveler.

But in the goods transportation space, that problem has been solved so long and so well that it's taken for granted—to the point that customers don't just expect two-day service anymore, or next-day service. Now they want same-day service. And when companies like Amazon promise such service, companies like UPS—along with its principal U.S. competitors, Federal Express, the post office, and the many smaller start-ups that have appeared to help fill that niche—have to make it happen. UPS was the first company to solve the door-to-door riddle, and they are now the biggest.

UPS traces its roots back to a small messenger company founded in 1907 in the basement of a Seattle saloon by a teenage errand boy named Jim Casey. After a few years of running errands, Casey and several partners switched the model to delivering packages, primarily business to business. The idea slowly spread from coast to coast. Later it expanded to include business-to-consumer deliveries, first by shipping items ordered from such outlets as the Sears Catalog and its rivals—the analog precursors of e-commerce—and now for dot-com retailers, which represent almost half of all UPS deliveries as of 2015.

This shift has been a difficult transition for UPS and its competitors, because instead of hauling ten or twenty or a hundred boxes from a manufacturer to a store—the mainstay of the delivery business for most of the twentieth century—the consumer space for the most part consists of one package to one house. That means many more stops and many more miles for essentially the same earnings. Solving the last-mile problem one house at a time is an expensive proposition.

But that costly yet inevitable shift, surprisingly, is not what keeps Noel Massie up at night. And his desire to explain that, as well as his company's role in the door-to-door universe, is what brought me to his office in the old Olympic Building.

Massie guards his time as carefully as ORION crafts a delivery route: the business of minutes thing carries over to his personal calendar in a big way. His personal schedule is so packed, we had to book our meeting sixty days out. "Don't feel bad," Massie tells me. "Mayors and city councilmen are handled the same, if they get a meeting at all." He grins at the thought of saying no to the mayor of LA. "Seriously, I'd rather meet with students."

Each day of the week on Massie's calendar is fully purposed: Monday is dedicated to sales—revenue, customer acquisition, marketing, volume, what he calls the "Where are we?" meetings. Tuesday is reserved for operations: performance, costs, efficiency, error rates, the "Are we hitting our goals?" meetings. Wednesday is set aside for customer visits—he has 144,000 regulars who ship with him daily, and he'll pitch sought-after customers directly to entice them into the fold. Thursday is for "externals"—the day Massie devotes to local organizations, charities, schools, community engagement. He's active with the Urban League, the United Way, and the Los Angeles Chamber of Commerce, where he became chamber president in 2014—the first African-American to hold that post in the organization's 125-year history. On Fridays he wraps up the week with intensive one-on-one meetings with division heads, directors, and anyone that doesn't fit in the other more categorized days. Once a month he holds his staff meeting outside the offices, assembling instead at a different local nonprofit around town. In return, he offers his host UPS's help in logistics, shipping, online presence—whatever a group needs. After one of these "outside-the-box" staff meetings, Massie was intrigued by his host for the day, a group called Trash for Teaching, which rescues and repurposes overruns, seconds, discontinued items, and other useful "waste" from businesses and manufacturers for use in science, technology, and art classes at schools throughout the region. He returned with three hundred UPS volunteers to reorganize and redesign the offices, warehouse, and distribution system at this LA nonprofit.

It's when he is out in the community, talking to nonprofits, to schoolkids, to meetings at the chamber of commerce, that Massie expresses his great fear for the future of his business and the nation's economy. This is what keeps him up at night: he is worried that the day is approaching when his trucks won't be able to complete that last mile on time. Or at all. And it will not be due to any failing on the company's part.

"My business is mostly about the truck. Because the last mile in the life of every product in America happens in a truck. The glasses on your face, the tie you're wearing, the phone in your pocket. It may get here in a container. It may spend time on a train. It may fly in a plane. But the last mile is always in a truck. Unless we go back to horses and buggies, or someone invents teleporters, trucks are going to be what we use for a very long time. At the end of the day, trucks are the most important vehicles on our highways."

He is leaning forward at his desk at this point, pausing for effect before revealing his main concern: "What do trucks need? They need roads. They need infrastructure. They need to be able to go where they need to go. And we are already far past crisis when it comes to infrastructure investment in this country."

He ticks off the problems that keep him up at night: failing bridges, potholed streets, congested ports, endless traffic jams. Truckers on overnight hauls can't even find safe parking half the time. As vital as trucks are to the economy and our way of life, Massie says, they are treated like interlopers on America's roads. He'd like to see dedicated highway freight lanes—high-speed lanes just for trucks, isolated from passenger traffic—and greater public transportation investment to take cars off the road, making room for those freight lanes and more trucks. This is not an idle wish: demand for goods delivery is going to double in the next twenty years, he says, and if our infrastructure doesn't keep pace, what will happen then?

"It's simple, really. Trucks are like the bloodstream in the human body. They carry all the nutrients a body needs in order to be healthy. If your blood stops flowing, you would die. If trucks stop moving, the economy would die. That's not hyperbole. That's not embellishment. That's just math. And yet—and this is what really gets me—the general public hates trucks. People have become truck haters. They want them off the road. They oppose improvements that would keep the economy moving and growing. It's already hurting our business. People don't know what they're asking for. They would paralyze America if they had their way."

Sometimes it seems the paralysis Massie fears has already arrived. All it takes to see it is a drive on what is arguably America's most important highway.

California's Interstate 710 is unique: it was conceived as the first highway built primarily for trucks. What else could a freeway that terminated at a commercial seaport be for? As early as the 1920s, city planners and harbor investors began financing a road to connect the ports of Long Beach and Los Angeles to what was then the world's largest master-planned industrial district, a field of factories purposely aggregated south of downtown LA, populated by General Motors, Chrysler, Studebaker, and a dozen other iconic brands of the time. The idea was to create a direct north–south highway conduit for American-made products to be shipped out of the ports to the rest of the world.

It took decades more for the conduit to be built, reimagined, extended, and transformed from a locally sponsored road to a state highway and finally incorporated into the Interstate Highway System in 1983. This triumph came just in time for the containerization revolution that forever transformed the movement of goods and the direction they would flow. The road that was intended to foster a flood of exports from the U.S. instead enabled an era of unbridled outsourcing and imports flowing to the U.S., although it still served its original purpose of connecting the ports to the rest of the country.

This it did brilliantly. Too brilliantly. The 710 became the single most vital highway for consumer goods in the country. And then the age-old problem of induced demand and eventual overload wreaked its inevitable havoc. By 2015 the 710 at rush hour had become a morass. Bumper-to-bumper big rigs fill the lanes for miles, belching diesel fumes and slowing traffic—cars and trucks alike—to a crawl. The smog is a horror for the neighborhoods that the freeway traverses. The health impacts—childhood asthma, respiratory illnesses—are measurable. Delivery times are delayed by the congestion, costing shippers millions in fuel and lost business. The constant pounding of all that heavy truck traffic, far beyond the design parameters of the roadbed, have broken the highway surface, requiring massive and costly repairs. The 1950s on-ramps, mere stubs by modern standards, fill up at peak hours, with traffic backed up onto surface streets, clogging them, too.

The most important freeway in America has turned to quicksand.

Two persistent problems now affect this primary link to the nation's busiest port. There is an obvious need to increase the freeway's capacity. And there is a separate but related need to actually *finish* the freeway, which for decades has fallen just over four miles short of where it's supposed to go—a failure that has consequences for traffic flow throughout the region.

Both problems will require billions of dollars to fix. Both are controversial. And both have aroused the wrath of environmentalists, legitimately aggrieved communities that border this freeway, and what Massie would call the "truck haters."

The capacity problem is less controversial. Only the nature

of the fix is in question. At peak periods, the 43,000 daily truck trips to and from the port already overload the freeway. It ranks among the worst in the nation for congestion delays and also has one of the highest rates of big-rig accidents in the state. By 2035, there will be an estimated 80,000 daily truck trips crammed onto the same 710—nearly doubling the number of big rigs vying for space on the six-lane highway, a recipe for paralysis, pollution, and door-to-door disaster.

Then there's the gap. The freeway runs for twenty-three miles from the port, then just stops in the middle of Pasadena. The last leg of the freeway was supposed to run through the city and the adjacent (and very affluent) town of South Pasadena, dumping the truck traffic on the Foothill Freeway, an east–west corridor that connects directly to the Inland Empire's rail yards, distribution centers, and goods-movement nexus. What's good for goods, however, may not be good for local communities. Many neighborhoods carved up by the original 710 construction lacked the political and financial clout to oppose the project or to demand concessions, but the Pasadena area is another story. Community groups there have blocked all efforts to close the gap for decades.

But the approaching tidal wave of demand and the impending paralysis that Noel Massie and other business leaders fear have forced the state's hand. In 2015, two sets of proposals to fix the 710 were put forward.

State and local highway authorities have floated a couple alternatives to expand the capacity of the existing freeway: a $4 billion plan to add two new lanes, one in each direction, along with bike and pedestrian walkways that play on the nearby Los Angeles River, which the freeway parallels. Or there is a more costly $8 billion proposal that would be Massie's dream come true: four new elevated lanes for freight carriers only. The trucks would be separated from the cars and, theoretically, everyone would be happy.

One additional proposal put forward by the ports and local groups tired of choking on pollution would add electric power lines overhead so that zero-emission electric trucks could traverse the 710 corridor, then switch to battery power when leaving the freeway. And all of the plans will require much cleaner trucks than the current generation of diesel big rigs, as state and federal law demands sharp improvements in Southern California's notoriously poor air quality. Trucks in California will have to slash emissions by 90 percent by 2030. Startups, universities, retailers, and established truck manufacturers are joining forces in a race to develop next-gen big-rig truck technology that can match the power of diesel engines without the deadly emissions and the obscene gulping of fuel (the trucking industry powered through $147 billion in fuel in 2014). Twenty-two companies are working together on one such promising superlight experimental big rig called the WAVE—for Walmart Advanced Vehicle Experience—that uses a hybrid system consisting of a powerful battery electric motor coupled with a micro-turbine engine that together can cut emissions and fuel use by up to 241 percent. But a commercially viable version of the WAVE (that is, one that's cheap enough) may be a decade or more off, if it's even achievable at all.

Absent such a paradigm-shifting technological advance actually hitting the road soon and in large numbers, community opposition to any proposal that would allow more trucks or increase the freeway's footprint has already formed. Whatever alternative is selected, some sort of legal battle, and the associated years of delays, are inevitable. And that doesn't even begin to address the question of where the money will come from to build any iteration of the project.

As for the 4.5-mile gap, the state has put forward multiple proposals to complete the 710, including closing the gap with a $5 billion double-decker tunnel to accommodate freight traffic, or focusing solely on passengers with new light rail or rapid bus routes that would close the gap for human passengers but not goods movement. Of all the proposals, only the tunnels would relieve traffic congestion throughout the region by providing another direct route between the port and the warehouse zone of the Inland Empire. But community opposition to any closure of the gap is vociferous. Why, members of the affected communities ask, should they have to live with a massive road improvement that will primarily benefit businesses outside their community? Why should they pay the price and the businesses reap the profits?

And there it stands for now: the most important goods corridor in America versus the legitimate concerns of communities who fear the shoring up of vital infrastructure.

This is very likely the last big freeway project in California. There is no room for any new freeways, and no money if there were room. As for expansion of existing roads, there is no need greater than the 710's, a freeway whose fate affects not just local communities but the entire nation and the goods-movement system itself. So far, no solution on the table is able to satisfy the concerns of opponents while also relieving the overload already present, much less looming in the future. It's a stalemate.

"I don't know if it's a cultural thing in America that people feel entitled to the cement and the roads without having to pay for them, without having to understand how the system works, or that our economy depends on it continuing to work," says Noel Massie. "It's clear we need a healthy ecosystem that allows the movement of goods, of food, of everything in the future. It's not clear that we're going to get it."

And yet, that infrastructure (a word, by the way, that writers try to avoid because of its power to make eyes glaze over) that no one wants to pay for is still working miracles.

One such miracle is standing on an open space adjacent to a taxiway at Ontario International Airport, an air shipping container just in from Hong Kong awaiting the next flight out to the East Coast. The container is watched by an armed guard standing by, and is under constant live video surveillance as well. The container is piled high with small cardboard boxes on which no brand marking is visible, although everyone knows what's in them. The contents are worth more than $3 million, which explains the unusual security measures.

This is how the latest model of the Apple iPhone gets from the factory to your home. The black market is just too lucrative to take chances. When phone-making rival Samsung fetches its new phones by truck, it sometimes goes even further, sending decoys out to fool and possibly catch organized phone thieves.

Welcome to UPS's regional air hub in Ontario, California, the crown jewel in Noel Massie's empire—and one of the big potential beneficiaries of an expanded 710 freeway. In many ways, it serves as the air version of the Los Angeles and Long Beach port complex, a vital link in the movement of goods between America and Asia.

One of six regional UPS air hubs in the continental U.S., Ontario is strategically located in the goods-movement capital of the West, the freeway-rail-air desert nexus of wide-open space and cheap real estate known as the Inland Empire. Here the warehouses seem to stretch to the horizon, a vista that boosters call a vital source of jobs and critics decry as a wasteland of pollution and poor pay. Most major retailers and manufacturers, from Walmart to Skechers, maintain enormous distribution facilities here, with freight coming to them from the ports, then moving out again through UPS and a host of rival carriers. As a result, Ontario has also become UPS's largest ground operation as well as an air hub.

This is where the smartphones bound for America from Asia first come. This is where planeloads of perishable fresh flowers land one after another in time for Mother's Day. This is where the ancient Terra Cotta Warriors museum exhibit first landed from China. Ontario is a place of constant motion and constant cargo both exotic and mundane—an installation where, in one form or another, everything a delivery company can do is on display.

UPS has 539 cargo jets flying nearly 2,000 trips to 728 airports worldwide every twenty-four hours. (Federal Express has an even bigger air presence, but UPS's ground operations dwarf its chief competitor, making it the larger carrier overall: $11 billion more in annual revenues than FedEx's $47.1 billion.) About twenty UPS flights arrive and twenty depart every day at Ontario, with the first touching down at 4:00 a.m. with overnight deliveries to all over Southern California. Because time is so short, the containers on this plane are presorted so they can be loaded directly onto feeder trucks without cracking them open—one truck for San Diego, another for downtown LA, another for the South Bay, or wherever the parcels are bound. The entire plane load—thirty-eight containers; more if it's a 747 jumbo jet—are out of the airport and on the freeway in less than an hour. "In LA traffic, you're sunk if you're not on the freeway by 5:00 a.m.," says Don Chubbuck, hub planner and industrial engineer at Ontario. "That's why we do the sort on the other end." The overnights will get to the local UPS operating facilities by 7:00 a.m., 7:30 at the latest, for a quick load onto brown box trucks in time for delivery that day.

Air shipping containers look nothing like the big steel cans used on oceangoing cargo ships. Air cargo cans are made of lightweight plastic, portions of which are transparent, and are shaped into semicircular cylinders with flat bottoms—mirroring the shape of a stripped-down jetliner's interior. UPS jets are basically passenger airliners stripped of seats, carpeting, overhead bins, and every other accouterment other than seats for the pilots and a very tiny airplane galley for snacks and drinks. The echoing cargo area has rollers embedded in the floor so the containers can be pushed quickly into place without need of heavy equipment. When unloading, crew members roll them to the cargo hatch, which is much wider than the doors on passenger jets. Then a motorized contraption called a K Loader, which looks like a gigantic scissors jack used to change a flat tire, reaches up to the two-story-high hatch and grabs two containers at once. An entire aircraft with 200,000 pounds of cargo can be unloaded in minutes this way.

Except for the direct loading of the presorted 4:00 a.m. overnights, the air containers are trucked into a 750,000-square-foot hub center adjacent to the runways, where they are lined up next to banks of massive conveyor belts. At the time of this writing, the containers were unloaded by hand onto the belts, but an automated facility to do this robotically was being set up. The belts carry the packages past overhead scanners that read their delivery bar codes and route each one to the proper trucks, which are backed up flush to the opposite side of the conveyor belts. Rivers of packages are sorted to their destinations this way at high speed. International goods are first diverted to a special U.S. Customs post that's manned in the late-night and early-morning hours before they, too, hit the loading conveyors.

Multiple flights of lightweight, high-value products from electronics to jewelry fly in daily from Asia, usually via Anchorage for refueling. The incoming planeloads are unloaded at Ontario if the contents are bound for one of the seventeen Western states. The rest continue on to UPS's main "Worldport" air hub at Louisville, Kentucky, where the sorting facility is big enough to house eighty football fields and can ship more than 400,000 packages an hour.

On the other side of the Ontario hub, incoming ground-shipped packages from homes, businesses, and the Inland Empire's hundreds of product distribution centers—amazon.com, target.com, nordstrom.com, macys.com; the list goes on—are disgorged into a football-field-sized sorting area. Teams of sorters pull goods from fifty enormous aluminum chutes, each tagged for a different part of the country. Once sorted, those packages are placed on feeder trucks bound for other Western states, or shorter-haul trucks to one of the local railheads for transcontinental rail shipment.

A parallel operation—the "small sort"—gathers documents and small packages bound for the same destinations into large bags that are scanned as single parcels and sent on their way (bagging them makes for more efficient handling and less damage). A third sort—"the irregulars"—uses trains of motorized dollies moving through the hub to deliver to the proper outbound trucks such large or oddly shaped items as tires, machine parts, heavy crates, works of art, and refrigerated containers carrying perishable prescription drugs or medical samples. And yet another area—a long line of truck bays facing away from the runways—is filled with brown box vans being loaded overnight with package deliveries to local homes and businesses.

The single most common product type shipped by UPS is—perhaps unsurprisingly in this age of e-commerce—the consumer retail product category. The next most common in order are car parts, medical supplies and drugs, professional services (mostly legal and real estate documents), and industrial supplies and products. UPS workers at Ontario see the economy in real time: which companies are shipping more stuff, which products are moving slowly or exploding in popularity. Massie keeps careful track of such data, but guards this proprietary information with a level of secrecy appropriate to the confessional—an apt metaphor, because his customers, from small businesses to Fortune 500 titans, give him critical information they share with no one else, and that their competitors would love to possess.

In such a ceaseless hive of goods movement, there are occasional errors—"miss-sorts," UPS calls them. This is what causes a package to be lost or late or sent to the wrong destination. It is something Chubbuck and his industrial engineer colleagues at Ontario have been working on since he arrived there in 2007, when the error rate at the hub was one in 250. That works out to a 99.6 percent accuracy rate, which sounds good until you apply it to millions of packages. Then it's 4,000 misdirected packages per million, and Chubbuck calls that "completely unacceptable." In 2015 he said some of the sorting belts in the Ontario sort achieved one in 20,000 miss-sorts, a vast improvement—a 99.95 percent accuracy rate (50 misdirects per million packages). But Chubbuck says that is *still* too many errors, and he and his colleagues are working to bring the rates down even further. "It is fine for shipping candy bars—maybe. But how about for medical? When somebody needs X-rays for surgery, one out of 20,000 just isn't good enough." And so the work on errors continues.

Noel Massie believes that automation will ultimately solve, or at least manage, such problems as sorting errors, just as it has helped with route optimization. Robotic aircraft—jetliner drones, in other words—will be coming soon, he says. If you ask the pilots, they'll tell you the robot planes are already here in all but name. UPS's cargo planes can automatically approach the airport, lower the landing gear, land the plane, taxi off the runway, and then park and shut down the engines—all without the pilots touching a single control. In fact, when weather and visibility are poor, it's the *only* way to land the plane. Driverless trucks will be next, Massie says; if not the delivery vans, then certainly the feeder trucks plying the freeways. Safety, efficiency, lower cost, and the ability of trucks to drive nonstop without the need for sleep make the rise of the robots inevitable. But none of that is going to help him sleep soundly at night, he says.

"Technology's not the problem. Traffic's not the problem. The problem is infrastructure. That's what's going to stop us."

Many people will shrug it off as just traffic jams and inconvenience, he says, but the view from Massie's office on the fifth floor of the Olympic Building reveals more. If overload causes the delivery drivers of UPS to take a mere ten minutes longer on the average route, the company's costs go up $125 million. That means cost goes up for *everybody* who sells and buys stuff. And then multiply that a thousandfold, or ten thousand–fold, for every other business that has to move things door to door, which is virtually every business. Massie can feel the pressure of those mounting minutes gathering, an inexorable glacier of time. "We are already overloaded. And the fact is that the world is going to go from 7 billion people to 9 billion people by 2050. That's a million more people a week, every week for thirty-five years. If we don't have infrastructure keeping pace with that, what's our world going to look like?"

Massie sighs. He is on his soapbox a bit, but trucks and goods movement are his fascination and his livelihood. He is a problem solver in his own shop, and it frustrates him to see the larger problems that he cannot control hurting his business and his country. He finishes his thought: "And just to be clear, we are *not* keeping pace."

Then he glances at the clock, and the minutes passing, ready to move on to the next cycle.

Chapter 11

PEAK TRAVEL

Traffic engineer Ted Trepanier's career-altering epiphany hit him like a ten-car pileup that long weekend of July 15, 2011, better known in Los Angeles as Carmageddon. That's when he finally noticed America's transportation revolution hidden in plain sight.

Not that Carmageddon's planners sought revolution. All they wanted was to fix the soul-killing traffic on Interstate 405, and to do it the old-fashioned way: by adding lanes, the go-to gridlock solution for generations.

Trepanier watched with interest from his home in Seattle, then incredulity, as Carmageddon turned into Carmaheaven and the worrisome lane closures made traffic, however briefly, better. Traffic planners had bedrock beliefs about how to fix things and how things worked on streets and highways. Now, in Trepanier's view, those assumptions went out the window.

Ted Trepanier had embraced a vision handed down long before he was born, a vision of ever more cars, trucks, lanes, and parking spaces. Roads and streets were, first and foremost, for drivers, and their free movement was everything. Such car-centrism had been wildly controversial a century ago but has since taken on the staying power of Moses's tablets. That's how the world worked. Until now.

His colleagues viewed Carmageddon as a curiosity, but Trepanier saw a turning point—a shift from a world dominated by car ownership to one of mobility by any means. Just a few years ago, Carmageddon might have been more like the nightmare everyone had feared, but something fundamental had changed, and the bewildering trends Trepanier and his colleagues had been trying to ignore since the start of the new century now made sense. That's why app-enabled car and ridesharing were displacing traditional cab and car rental companies (who were resorting to lawsuits and lobbying to stave off the future) and why carmakers and computer companies were investing billions in perfecting driverless cars that were best suited to a sharing economy, not a three-cars-for-every-household world that Detroit longed to foster. That's why the country had hit peak cars, peak shopping trips, and peak roads years before while online giant Amazon was in growth mode, homing in on robot warehouses, 3-D printers, drones, and that Holy Grail of online shopping, same-day delivery. Why drive to the store when someone will drive your purchases to you? Next-gen Amazon computers have even begun predicting customers' orders and shipping *before* the orders are actually placed—all to make personal driving unnecessary, even unattractive.

And perhaps most important this switch from a focus on car ownership to an embrace of a smorgasbord of mobility options that played out during Carmageddon explains why so many Millennials, America's largest demographic, are losing interest in driving and buying cars. The number of "zero-car families" has been steadily climbing every year since 2007, after shrinking year after year with dependable regularity since 1960. One in ten American families has no car at all. Detroit would have sold another half million cars in the past decade if not for car-sharing smartphone apps. The transportation world had been passing him by, Trepanier realized, and, like all revolutions, the new trends could be seen either as dire threat or golden opportunity.

Trepanier, a baby boomer who grew up believing that his car's pink slip was the key to freedom, suddenly understood his twenty-seven-year-old son who has never owned nor wanted a car, who sees car ownership as a cost and burden and a parking nightmare that limits rather than unleashes his mobility. His son just wanted to get door to door simply, easily, cheaply. "Why would I want to blow all my money on the care and feeding of a car, Dad?" he had once said. "So I can pay to park it all day at work, or outside places I can get to just fine without a car?" Carmageddon didn't surprise Ted's son at all. From his point of view, LA had simply, finally wised up.

And so, says Trepanier, had he. He left the government and joined Inrix, a Microsoft spin-off that created a free smartphone app that gathers real-time data on traffic worldwide better than all the billions spent by government on sensors and traffic cameras. Inrix is a major player in crowdsourced, cloud-stored traffic mapping that it uses to enable mobility rather than just offer GPS driving directions. BMW is deploying it first, as is the New Jersey Department of Transportation. Its technology tells drivers how to get places not based on the shortest distance but on real-time traffic, congestion, construction, public transit, and parking data. Your BMW will be able to tell you to pull off the freeway at the next park-and-ride and hop on an approaching commuter train or bus because it will get you where you want to go faster and without parking hassles and the congestion that results from drivers circling in search of parking. Carmakers are actually building technology that tells people to do the unthinkable: *to get out of their cars.*

Monorails. Flying cars. Nuclear-powered cars. A helicopter in every garage. Subway bullet trains traversing the country. Moving sidewalks. Magnetic highways to guide vehicles so drivers can relax and play board games with the kids. Rocket planes that go suborbital to cover long distances quicker.

We were supposed to have all these by now, or so the predictions of the future went a couple generations ago. Traffic was supposed to have been solved. Energy and pollution, too. All you had to do was go to the World's Fair back in 1964 to find that out.

The problem with predicting the future is that what seems to make sense today often doesn't match up with future reality. Computer networks and the Web were imagined years before they were part of daily life, but no one predicted how networked smartphones would impact transportation, enabling ridesharing, traffic avoidance, and e-commerce that displaces trips to the store, the bank, and the post office. A hundred years ago, streetcars were predicted to last forever as America's most popular form of mobility because they were clean, efficient, cheap, and easy. Fifty years ago, the idea that bicyclists and pedestrians would want to reclaim lanes from cars would have sounded absurd. Or rather, having local governments allow and encourage it would have sounded preposterous. And no one would have predicted that doing so would improve traffic flow and safety. Yet efforts in New York, Los Angeles, San Francisco, Portland, and cities around the world have seen a reduction in congestion after they "calmed" traffic by forcing cars to slow down and share the space. Many drivers and civic leaders remain unconvinced by the data that supports the value of traffic calming, and the political battles over such ideas have become yet another front in the culture war. It is one of the perversities of the era that, in the political arena, bikes and trolleys are deemed "liberal," while cars are "conservative," even though in real life all political persuasions use these various modes of travel equally. Or that cars are by their very nature the least conserving form of personal transportation.

Surprisingly for a car-centric community, almost all the official traffic studies and plans for the Los Angeles freeways, going back to the 1920s and continuing on into the sixties, called for mass transit embedded in the freeway medians: heavy rail early on, then light rail, then lines to connect to and revitalize the existing Pacific Electric street car system, and finally, in a fit of Disneyland inspiration and envy, monorails with quiet rubber tires cruising up and down the centers of freeways. Many of these plans were supported by the Los Angeles Chamber of Commerce and other prominent business leaders. Yet all were eventually discarded, either by voters who rejected extra taxes, or by politicians angry that their city wasn't on some proposed route, or simply because the age of the motorcar had arrived and Los Angeles saw itself as ground zero. Let San Francisco build its sixties-era Bay Area Rapid Transit System trains to connect all the cities up there. LA wanted freeway lanes, not rails.

And so Los Angeles's privately owned trolley and interurban rail system—once the largest such system in the world—was shut down, the rights of way sold off. My streetcar suburb of Seal Beach marketed its real estate in the early 1900s with newspaper ads pointing out that Pacific Electric's Red Car Line of electric trolleys could get prospective buyers there from downtown LA for a nickel. Now the last Red Car is a rather forlorn single trolley parked on the last twenty feet of track next to the Seal Beach town library, converted into a small museum with scant hours and even more scant visitors. There are no more trolleys to the beach cities and ports of Southern California, which once delivered both passengers and freight in the region. The lines that served the areas of the city now occupied by Los Angeles International Airport disappeared, too, although air travelers today would give much to be able to take such a conveyance instead of the 405 Freeway. Meanwhile, San Francisco constructed what is now considered to be one of the more cost-effective mass transit systems in the U.S. built since World War II. Travelers can land at the airport there and take a BART train almost anywhere in the Bay Area.

Most of the nation followed LA's lead.

Not until 1990 did Los Angeles embark on an ambitious light rail and subway building project in an attempt to re-create some of the mass transit service so blithely abandoned a generation ago. The city could have kept all that for pennies on the dollars now being spent.

It's tempting to judge those earlier decisions harshly, to condemn the shuttering of a valuable transportation asset and the refusal to build a new one when it would have been so much easier and less expensive to lay those tracks when the freeways first went in, rather than trying to shoehorn them into a built-out urban landscape today. But were those decisions wrong? Mass transit ridership was dying in the region even before World War II. And for all the money being spent on new light rail and trolley systems now, ridership is only a fraction of what it was a century ago. Cars won. And the decisions made to reject those multimodal freeways were rational at the time. People wanted cars. They didn't want to see America from the train. They wanted to see the U.S.A. in their Chevrolets. They wanted to drive to work in air-conditioned comfort, not walk to the streetcar or train station, then wait around on crowded platforms.

All the billions spent on mass rail transit in LA in recent years, the most ambitious build-out of multiple routes anywhere in the country, has not reduced car traffic jams as hoped. It helps somewhat, but the reductions make it hard to justify the expense. This mirrors the experience nationwide, even as about 25 percent of surface transportation spending goes to fund mass transit. Mass transit use has picked up a bit in recent years but still is lower than it was a quarter century ago and far below its absolute peak in the

1920s, when it was the best and most desirable way to get around the nation's cities and suburbs. Indeed, suburban development followed the extension of mass transit lines back then in the era of streetcar suburbs, because the trolleys were considered a prerequisite for suburban development. Most streetcar suburbs have been absorbed into cities proper since then, and suburban development after World War II eschewed following mass transit and instead relied on car accessibility. The new mass transit spending is not enticing waves of new riders to abandon their cars and ease road traffic. The convenience of the car parked in front of the house trumps the inconvenience of getting to a train or trolley or bus. In the 1920s, Americans were not deterred by this last-mile problem. People walked to the stop—no big deal. Now it is a very big deal. We are conditioned to expect service door to door, and mass transit just doesn't do that.

Public transit has become yet another tussle in the culture wars. Critics say mass transit spending, at least on the federal end, should be shifted to highways in order to deal with traffic. But that strategy can provide no more relief for congestion and overload than the mass transit projects. The $3.6 trillion in unfunded maintenance and repairs on the nation's roads and bridges would barely be scratched by slashing transit spending. The hidden subsidies for drivers, who have never borne the full cost of the roads they use and who now barely pay half through gas taxes, are just too great. Besides, what would the money be used for? The strategy of adding new lanes—even boosting that highway in Texas to twenty-six lanes—just doesn't work. Traffic is worse on the 405 after the Carmageddon lane addition. It's been shown time and again.

Los Angeles and California have spent millions on traffic control centers with state-of-the-art monitors, street and freeway cameras, computer algorithms that let traffic lights and freeway on-ramp meters be timed and adjusted to alter and maximize the flow of traffic. In truth, these systems are amazing pieces of technology, wonderful in emergencies and for disaster response—and making sure celebrity limousines get to the Oscars on time. But they really aren't very good at "controlling" everyday traffic, at easing congestion significantly and keeping ordinary cars moving at ordinary times. Their effect is negligible on that score, perhaps shaving a couple minutes here or there off a trip. Traffic congestion in the city and region remains some of the worst in the country.

The old approaches, experts such as Trepanier say, just aren't working. The old predictions are failing. The old rules are wrong. Building for speed in dense urban landscapes—laying down stroads where there should be streets—kills people more than it eases congestion. And the old way of paying for roadbuilding and maintenance based on gasoline purchases isn't paying for nearly enough. The fact that Americans are driving less per capita than in years past has only exacerbated the chronic underfunding of transportation in America.

No one was predicting any of this ten years ago. No one.

The Millennial generation is, statistically, somewhat less interested in cars and driving and suburban living than older generations. How much of that is due to a bad economy and a tough job market is hard to say, and there is a tendency in the popular media to overstate the Millennials' divergence. Plenty of them still drive, plenty of them want cars, plenty of them would jump at a nice home in the suburbs if they could afford it. The divergence is real, however, and measurable: Americans between the ages of sixteen and thirty-four drove 23 percent fewer miles in 2009 than in 2001—a greater decline in driving than any other age group.[1]

That might sound dramatic, but focusing on it, using it to posit a longer-term shift that will drive transformation in transportation, is yet another risky prediction. And it distracts from a much larger and more far-reaching distinction: Millennials have embraced the app-driven sharing economy that really is disrupting the transportation world. It's calling Uber instead of driving home drunk, or relying on the crowdsourcing of the traffic app Waze. It's not about Millennials loving buses, bikes, or trolleys more than cars. The trends more suggest a tendency to be transportation omnivores in ways that car-centric older generations are not. Ride services such as Uber and Lyft are alluring because, with a simple app, a car can be summoned to take you where you need to be, without parking concerns, without car payments—other than the fare for that one trip—or insurance costs, or responsibility to navigate. If your road travel needs are below 10,000 miles—the average nationally is about 13,400—dumping your car and taking a ridesharing service may save you money. The inefficiencies of owning a car, of having it sit idle twenty-two hours a day even as it is Americans' second largest expense after housing, is especially grating on a generation that has been burdened with a flagging economy and shortages of jobs, not to mention unprecedented educational debt. It is as Ted Trepanier's son explained: Why would Millennials love cars? They're too damn expensive for what they bring to the table. Mobility agnosticism is the Millennial superpower that is truly disrupting transportation—that and the emergence of a few key technologies.

President Obama's secretary of transportation, Anthony Foxx, is hanging an entire national policy on this shift, a policy he has dubbed "Beyond Traffic." His report of the same name calls for a reordering of priorities more in line with the Millennials' mobility tendencies. Assuming the country gets over being stuck on an inadequate 1993 gas tax and makes transportation funding sustainable for the first time since Jimmy Carter was president, Foxx envisions making mobility, not cars, the center of a balanced approach to transportation policy. He says the country must embrace new technology, particularly automation and connected cars, for safety, congestion reduction, and efficiency. Continuing current policy will bring disaster within the next twenty-five years, according to Foxx, whose report provides a grim summary of the current transportation picture:

> In the race to build world-class transportation, America once set the pace. We used to have a big lead.
>
> In the 19th century, we built the Erie Canal and Transcontinental Railroad. In the last century, we took over building the Panama Canal, completed the Interstate Highway System, and set the world standard in freight transport and aviation.
>
> But our lead has slipped away. We are behind. Way behind. The quality of our roads, for example, is no longer rated No. 1. We're No. 16.
>
> And it is not just that our infrastructure is showing its age—our country, in many ways, has outgrown it. If you drive a car, you now spend, on average, the equivalent of five vacation days every year sitting in traffic. If you drive a truck, highway congestion has made you an expert at navigating bumpy side roads—and you are not alone. Every year, trucks are losing $27 billion on wasted time and fuel.[2]

Foxx's report is understated and diplomatic, 316 pages of reasoned argument, history, and prediction. But it accurately describes a system overloaded and failing, with grave consequences for safety, the movement of people, *and* the movement of goods that sustains our economy. This is not a cheery picture. It is a picture of a transportation system still capable of delivering incredible results that make life and work better and more prosperous, that takes us and our stuff door to door better than any other civilization has dared to imagine. But it's also a system living on borrowed time and showing the strain. It is a system that has become unsustainable.

The good news: there are five important trends that will bring change, in some cases massive change—if they continue.

Three of them affect the goods-movement side of the door-to-door machine. First, there is the transformation of China from a low-wage factory sweatshop economy into a true economic powerhouse, where workers are demanding—and getting—better wages, benefits, and working conditions. The ready supply of rural peasants who flocked from the countryside to keep China's factories moving has begun to dry up. Higher wages in China make offshoring of jobs and manufacturing less attractive to American companies. They have also roiled China's economy and relentless growth. This is driving the second trend, still nascent, of "re-shoring"—a resurgence of manufacturing in the U.S. that would have been offshored in the recent past. The more likely scenario is not a large-scale return of existing jobs to America but the increased likelihood that new technologies and products that emerge and create jobs in the future can stay in the U.S., sucking miles out of the door-to-door system for the first time in many years.

The third trend, also related to the other two, is the emergence of the fledgling 3-D printing industry as one of the major disruptive technologies on the horizon.[3] The promise of this "factory-in-a-box" technology is the ability to manufacture products locally at a competitive cost, even in small quantities. The latest 3-D printers are being used to produce increasingly complex products, from aerial drones to prosthetic limbs to car bodies. It may take a decade or more of development for the technology to mature sufficiently, but the possibilities are revolutionary. Imagine ordering a product online and instead of it being physically shipped from factory to store to customer, the proprietary design is shipped as a digital file via the Internet to your neighborhood 3-D print shop, where you pick up your purchase later that day once the printing is complete—transportation virtually removed from the process, "streaming" products as we currently stream video.

The technology is not there yet, but it's hard to imagine a greater disruption in the door-to-door universe if the range of materials it can fashion becomes broad enough, and the costs low enough. Such a technology would revolutionize goods movement . . . by eliminating the movement. And the overload.

The other two emerging trends are more about moving people than goods, although they could have profound implications for the have-it-now e-commerce economy as well. The first is the sharing/crowdsourced world of traffic apps and ridesharing. They are disruptive in their own right, demonstrating alternatives to car ownership never imagined before the rise of smartphones. But their greatest impact may be as a test case and as a necessary first step on the way to the most disruptive transportation trend on the horizon, the one that could inspire as much change as the invention of the car itself: the driverless car.

Chapter 12

ROBOTS IN PARADISE

The most startling thing about riding around in a driverless car is just how ordinary, even boring, it can be—once the shock passes of seeing a steering wheel move by itself. It takes about a minute for my brain to turn self-steering from stunning conjurer's trick into something commonplace. That's when I notice the other standout feature of driverless cars: everybody else on the road is passing us by. One vehicle zooms close up from behind, the classic sign of an impatient driver ticked off at the slowpoke in front. Once it becomes clear that the autonomous car won't be intimidated by the tailgater into speeding up, the car behind swerves into the passing lane and zips by. A human driver would probably take this affront personally. The robot car? Not so much.

The reason why all the other cars are passing the Google driverless car—or rather, the self-driving car, to use the search company's preferred nomenclature—is simple. It's the only car on the road obeying the speed limit. And obeying the law is one of those things it does scrupulously, which will eventually raise an interesting safety issue down the line. Another thing it does—rather jarringly, given its usual non-daredevil driving style—is react like a cat.

I saw this behavioral change happen halfway into my ride, as we were driving down a tree-shaded street that cuts through the

Google campus in Mountain View, California. All the buildings on this street house Google offices, labs, meeting rooms, and cafeterias where the food and coffee are shamelessly tasty, but the feel is more college campus than tech titan. Amid a steady stream of pedestrians and cyclists, the Google car pokes, smooth as butter, laser turret on top spinning and watching. I'm chatting with the test driver and her partner, who is tapping on a laptop displaying the stylized virtual reality of differently colored moving squares, circles, and vectors that represent the objects the car "sees." Suddenly the robot driver jams on the brakes. Did I say driverless cars are boring? I look up and see a bearded fellow walking in front of the now stopped car. He is balancing his own full-sized laptop on one palm and forearm while typing away with the other hand, staring at the screen intently as he jaywalks right in front of the car. He finally looks up, vaguely surprised to see the Google's Lexus SUV test car with the sensor array bolted on top, then gives a little wave before resuming his diagonal course across the street, back to typing and staring at his screen. Coders. They're all over the campus, it seems, focused on task, oblivious to environment. I can't help but think that a human driving the same street, maybe distracted by a conversation or taking a quick peek at a cell phone or just leaning over to tune the radio, could have been slower to react than the robot car, and the result could have been very different. Flying laptop. Flying nerd. Ambulance rides. Or worse.

Welcome to the brave new world of autonomous vehicles, where rules are followed, at least by the nonhumans, and where, if Google has its way, nobody will ever die in a car crash again.

Can that happen? Absolutely. Will it happen? That's uncertain. But *should* it happen?

That's my question to Ron Medford, the former number two at the National Highway Traffic Safety Administration, now safety director for the Google Self-Driving Car Project. We speak after my ride, after the requisite marveling at the gee-whiz technology, after Google has shown off what this car can (and, just as important, won't) do. Can and should are very different things, after all, and Medford has been a car guy—a human-driven car guy—all his life. So I want to know: Why does he think we should all follow this path Google is pursuing, this odd yet compelling deviation from its core Internet-search gravy train?

Medford's response is adamant that we absolutely should. His explanation boils down to two points: If you (1) care about your kids, your spouse, your parents, your siblings, the elderly, the infirm, or the lives of innocents, and (2) if you ever saw a human driver do something stupid, dangerous, or deadly (which is to say, if you've spent more than five minutes on the road), then you want self-driving cars to happen.

"As soon as possible," he says. "As fast as we can."

Most major carmakers are working on driverless technology, along with Google, a few smaller car technology companies, the major automotive parts makers Continental, Bosch, Delphi, and Tesla (where chairman Elon Musk predicts that human driving will be illegal someday), the rideshare leader Uber, and possibly Apple, although that company, as usual, isn't talking. Almost all the automotive powers have large outposts in Silicon Valley now, a clear indication that the future of mobility will be guided by software, not hardware. Behind this extraordinary pursuit of new technology, two very different philosophies have emerged, one evolutionary and more comfortable for consumers and car lovers to accept, the other revolutionary and more disruptive, but with the bigger potential payoff.

The mainline carmakers' approach is to treat autonomy as a feature, something to be introduced gradually without eliminating the standard car-owning and driving experience. Their autonomous cars leave drivers very much in the loop. First there's highly adaptive cruise control now on the market that can automatically respond to highway conditions and brake to avoid collisions automatically—when drivers decide to switch it on (studies show many don't). Then there's parking assist, a real blessing for the parallel-parking challenged, one of the main reasons new drivers fail their driving tests (leading at least a dozen states to drop parallel parking from their tests). These technologies are already available in some cars. Next might be lane-following technology, so you could turn over control to the car while on a free-flowing freeway, but not in stop-and-go city traffic. It would be like a traditional car with an autopilot function, with the human driver expected to remain vigilant and ready to take over at any time. In fall 2015, Tesla transmitted an over-the-air update to its 2014 and later models that enabled such an autopilot feature, with warnings that the technology was still in its early phases and so drivers should keep their hands on the wheel and be prepared to take over at any time. The user reviews were mixed, because a system that requires constant supervision from the driver is more curiosity than useful tool. But it did hint at the potential of such technology.

Eventually, full autonomy on all types of streets and roads could be introduced by the big consumer carmakers, but with redundant controls for the human driver to use whenever he or she felt like driving. The closest analog would be jetliners with autopilots doing ordinary tasks but with human pilots supervising and in control at critical moments. Such cars are being tested in Europe, the U.S., and Asia. Volvo is using a ring road in Sweden's second largest city, Gothenberg, where it expects to test one hundred cars in this mold with volunteer car owners by 2017. The University of Michigan has built a thirty-two-acre fake city to use as a test bed for the driverless products of Ford, General Motors, Honda, Nissan, and Delphi for safely testing both city and highway driving by robots. And Bosch is road-testing something it calls the "highly automated" vehicle, which is driven manually to a highway, where the driver can either continue at the wheel or switch to autopilot. When the car reaches the exit, it asks the driver to resume control; if there is no response, the car pulls over and stops. This approach has another advantage: automated driving on limited-access freeways with divided traffic lanes, off-ramps instead of intersections, and no traffic signals poses a significantly simpler technical challenge than surface street driving. That level of autonomous technology is virtually off-the-shelf at this point, as Tesla proved with what amounted to a customer beta test.

Carmakers plan to pitch such systems as the best of both worlds, a car, much like today's cars, with a really cool set of high-tech options. These options might improve safety when engaged, make drive time more relaxed and productive, and get people used to the idea of automation without the potentially off-putting discomfort about giving up total control to a machine. No one would—or could—be forced into giving up driving in these scenarios. We'd still enjoy the "freedom" of driving—as well as the freedom of driving drunk, distracted, or over the speed limit. In this scenario, while it might be true that such technology will take drivers out of the loop someday, it would be many years in the future, so there's little point in pushing hard now. With 265 million cars, pickups, and other passenger vehicles on the road in the U.S., nothing changes overnight.

The Google approach is different: the company's goal is to perfect the technology, then bring it to market whole hog, with completely autonomous cars that don't even have steering wheels or gas pedals. Just a start/stop button and an interface to tell the car the desired destination. The faster that transition happens, Google suggests, the better off the world will be. They started with stock Lexus SUVs as test beds for the autonomous technology, but in mid–2015, they began road-testing their own design for little electric pod cars designed from the ground up for autonomous city driving. This approach dives into the discomfort of ceding control to a computer and pitches it as a new freedom: freedom from having to pay attention. So you want to text in your car? Get distracted? Doze off? Feel free! A major cause of death and destruction in regular cars—distraction—is a welcome feature in Google's driverless reality.

"We think ours is the correct approach," explains Medford, the former number two at the National Highway Traffic Safety Administration before coming to Google. The handoff from robot to human is potentially the most dangerous moment when cars are partially automated, he says, particularly if a person who has not been paying attention to driving conditions is asked to take control in an emergency, as some of the early partially automated systems require. Google is more interested in reinventing driving than in adding features to existing cars. If safety is the overriding goal, along with helping the disabled and the elderly who can't drive become mobile, then, Medford says, Google's way is the right way.

The safety gains of full and ubiquitous automation are clear. Robots don't drink and drive, or get distracted, or get drowsy at the wheel, or speed, or randomly cross the centerline, or blow through stop signs or red lights. The main causes of traffic death and destruction simply go away. Robots' reaction time is close to instantaneous. Their radars and lasers have pinpoint accuracy and allow them to "see" 360 degrees simultaneously, through hedges and leaves, to identify pedestrians approaching crosswalks. The most powerful autonomous car sensor is the lidar, which sounds like a type of radar but is a completely different animal. The spinning scanner emits bursts of illumination at a rate of nearly a million flashes per second, a flickering light invisible to human eyes that bounces off everything in the environment—people, cars, buildings, curbs, squirrels. The process is something like insanely fast photography, but the result is a kind of three-dimensional machine vision that allows the car to perceive its environment in incredible detail in real time, with measurement of objects and their height, width, depth, and distance accurate to the millimeter. Lidar makes it possible for the driverless car to calculate a collision course with a wayward oncoming vehicle when it is still comfortably distant, and shift course so smoothly that no one even notices. Nothing can sneak up on the Google Car. In a world full of human drivers doing stupid things, autonomous cars are the ultimate defensive drivers.

But when it's all robots on the road, then the need to defend lessens and the real magic begins. When driving with their autonomous brethren, these cars are capable of cruising bumper to bumper on the narrowest of lanes at high speeds without a hiccup, driving in tight formation like precision aircraft teams, never wavering. It's as if they are on invisible rails, steady as trains. This is what robots are good at: precision, predictability, consistency. That's why they can drive so much better than imprecise, unpredictable, inconsistent humans.

"This technology can prevent drunk driving. Distracted driving. Drowsy driving," says Medford, who spent his career in government advocating auto safety technology. "And there's the issue of accessibility, too."

Medford is a polished, reserved guy with a great poker face, a survivor of multiple White House regimes, but this is one time during our chat when he shows his emotional investment in the Google driverless project. He recounts the story of his neighbor in Bethesda, a woman in her eighties, long divorced, her children living in other cities. She depended on herself and did so magnificently, leading a full, active, and independent life. Her car was her freedom. An avid bridge player, she drove her less-independent friends over to play cards. She picked them up to go shopping. She drove them to attend events, making sure they, too, could get out and stay active. Then one day she was in a minor fender bender at the library. The police came and she received a ticket. A week later, she received a notice from the Department of Motor Vehicles that she would have to take a full driving test, including parallel parking, to retain her license. The worried woman begged Medford for help: she hadn't parallel parked for years, she confessed. So he did what he had done for his teenage daughter years before, taking his sweet neighbor to a parking lot, setting up cones, and watched her try, and try, and try to do it. But she no longer had the range of motion in her neck to look back over her shoulder and steer and park at the same time. It was torture for her. Finally she got out of the car, weeping, and said, "I can't do it anymore, Ron." And she handed Medford the keys.

"That was the end of her independence," he recounts quietly. "She stopped playing bridge. She stopped going out. She tried to use public transit, but she had a fall stepping down from a bus, and that was it. She ended up in a nursing home. It broke our hearts."

He says he's certain that autonomous car technology could have altered the course of his neighbor's life, and that the potential for helping the elderly, the blind, and the physically challenged with mobility and independence is enormous. "This could be so meaningful for a large portion of the senior population. That's a big part of why we're pursuing this."

It's also why Google is more interested in moving faster than the car industry, Medford says.

That's not the whole story, though. There is more than simple disagreement over design philosophies and the pace of transition to full automation at work here. This divide in approach is also about preserving a business model based on selling the maximum number of cars possible. Carmakers are investing in and researching autonomy in part as a defensive measure, because they know that, in some form, autonomous technology is inevitable. Someday it will be ubiquitous and, eventually, required. But if it is presented as a feature incorporated into otherwise traditional cars, then the current model of private car ownership, of two or three vehicles per average household in America, might continue. A slick autonomous driving option then becomes just another selling tactic, a point of competition that can enhance car sales. And by not requiring technology that makes a car driverless all the time, but just under certain conditions—on freeways with well-marked lanes, for instance—carmakers can take a less technically challenging route.

Google, which has no existing car business to protect, but sees endless potential for marketing its autonomous technology, has a very different vision. A car in which autonomy is not a feature but the essence of the vehicle, so that it *can't be* driven by humans whether they want to or not, changes everything. Yes, it brings to an end the carnage of car crashes, of 35,000 deaths and 1.5 million trips to the ER every year. Yes, it gives mobility and independence to the elderly and the infirm who might otherwise be housebound. Yes, it reduces drunk driving and distracted driving to historical curiosities. But there's so much more. There is the flip side of the fully autonomous car, the true differentiator between the carmakers' evolutionary approach and Google's attempt at revolution: the Google car can do stuff without any humans in the car at all.

The car that travels on its own can remedy each and every major problem facing the transportation system of systems and, along the way, end car ownership as we know it. That's why transportation scholar, author, and blogger David Levinson of the University of Minnesota has proclaimed that "autonomous vehicles appear to be the next profound transportation technology."[1] That is, autonomous cars could be to regular cars what those same regular cars were to horses more than a century ago. That's what Levinson foresees, and he is not alone.

Imagine this scenario: fully autonomous cars have become ubiquitous. This is doable sometime between 2030 and 2040 (with early adopters appearing by 2020). Human drivers will become little more than hobbyists, with car driving relegated to the same status as horseback riding: as recreation, not transportation. What does this transition do for the world? What might it look like?

Let's say you open an app on your smartphone and summon a driverless car to your house. You need to get downtown to attend an all-day conference that starts in two hours. The app consults the latest crowdsourced traffic data and informs you the car will pick you up forty-five minutes before the conference starts to ensure an on-time arrival. At the appointed time, your phone buzzes: the car is outside your house. It takes you to the conference site, its route selected based on current traffic data; it drops you at the curb and then takes off to pick up another passenger. During the ride you read and answered e-mail, browsed the news, made a couple of phone calls, played some Words With Friends, and booked a dinner reservation. Drive time has become productive time. Neither you nor the autonomous car has to worry about parking at the end of the trip—a process that, in the once congested downtown area, used to be both time-consuming and expensive. Space once set aside for parking cars is now used more productively, in the forms of protected bike lanes, outdoor cafés, open space, and mini-parks. At the end of your conference, you use your app to schedule a pickup to return you home, your transportation needs met for the day—no fuss, no muss.

In this scenario, traditional car ownership makes no sense. If that was your personal car taking you downtown, then you would have to worry about finding and paying for parking, the driverless car cruising for a spot just like a human. And the same inefficiency that afflicts regular cars would kick in: the car would sit idle and unproductive all day. Instead, in a driverless rideshare scenario, any one of a buffet of options are available to you in seeking driverless car service: you could subscribe to a car plan much like a phone plan, buying minutes of travel rather than minutes of voice calls. Or perhaps payment is calculated by the mile. Or maybe you subscribe through a monthly fee like a data plan, with tiers of miles allowed, choosing a plan to match your needs. A thousand miles a month? Fifteen hundred? More? Or you might purchase any number of possible subscription plans, time-share plans, or contracts for autonomous vehicles on demand, with rates and terms kept reasonable by competition between providers. The key point here is that the car would belong to someone else, which might be a rideshare service, a car rental company, or the carmakers themselves, with customers paying only for what they need and use. Venues like Disneyland and Dodger Stadium and the Las Vegas strip might bundle tickets and hotel rooms with driverless service. Savvy local transit companies might get into the game, using driverless cars to solve the last-mile problem and get more people onto LA Metro's light rail or the LA–San Francisco bullet train now under construction. Would you use mass transit more if a driverless car whisked you to the station just in time to board—and it saved you money as well as time? New York City's transit company or Boston's or Portland's might enter such a market, simultaneously finding riders and easing traffic in their communities. They could offer multi-modal mileage plans, with savings for riders who take the bulk of their journeys on high-speed mass transit. Indeed, in this future scenario, carpool lanes could be converted to automated bus lanes; multiple driverless buses could use unerring robotic guidance systems to cruise bumper to bumper like trains at 150 miles an hour, drafting behind one another like race car drivers to cut wind drag and delivering passengers in record time.

For consumers, there would be no insurance costs for driverless cars in this scenario, no fuel costs, and no parking costs. What it would most resemble is a ridesharing service without the human behind the wheel, which would greatly lower the cost. And now you know why Uber, the global rideshare leader, which went from zero to ubiquitous since 2009, has hired away university researchers by the dozen and launched its own driverless car project. Robot Uber is on the drawing board. The company sees such a transformation of car culture as the inevitable outcome of a perfected driverless car.

This is why carmakers want autonomy to be evolutionary, not revolutionary, to keep that Uber robot at bay as long as possible. A world of fully autonomous cars à la Google or Uber would end car ownership as we know it.

It also ends cars as we know them. Think about it. The most common method of commuting to work in America—76 percent of the work trips we take —consists of one person in a car that could seat five or six people, that has an enormous trunk and a fuel tank with a range of hundreds of miles, all for a trip that, for the average American, is under fourteen miles each way.[2] Indeed, the average car in American cities travels a total of 36.5 miles a day for all purposes, while the average rural-based car clocks in at 48.6 miles all day. As car owners, we want vehicles that can carry a lot of stuff and people long distances, even though we usually don't do so, just so we're covered when we need that capacity. Nobody wants too little car to do what they need to do. But that's another built-in inefficiency of the current car-ownership economy. Not only are our cars parked most of the time, but they offer much more capacity than we need most of the time.

In the app-driven, on-demand scenario of driverless cars, the fleets for rent can be made up of a variety of vehicles purpose-built for each type of trip. Providers would build fleets out of a variety of vehicle types and sizes. Freed from the need to build cars that serve as general-purpose-vehicle versions of the Swiss Army knife, car designers could unleash a torrent of new designs for specific purposes: small one- or two-seat minis for short city trips; sleek in-line models for families of four seated one behind another for quick journeys to the beach; car bodies designed to easily store bicycles or surfboards or that have cargo space for trips to Costco. Robot shared vehicles could also usher in the rise of electric cars because range would no longer be an issue for most trips. Each car would serve multiple users in a day—instead of being parked twenty-two hours out of the day—on short trips with ample opportunities to recharge as needed between customers. The hundred-mile ranges of today's electrics could be easily outstripped with lighter vehicles. Larger autonomous cars capable of carrying more people and stuff or going longer distances—hybrids or fuel-cell powered, perhaps—would be available as needed. Gasoline-powered cars in the consumer space would be dethroned.

In this scenario, fewer cars would be needed nationwide. Far fewer. And the drivers that would exist would flow in harmony instead of battling to pass and cut off and wave fists at one another. Congestion and traffic jams would be a bad memory; no more new lanes would have to be constructed, and those we have wouldn't need to be twelve feet wide to accommodate human error. All those parking lots could become green spaces. We'd end up with smaller freeways, smaller streets, and much smaller transportation budgets for cities, states, and the federal government. These dollars could be redirected to maintaining and repairing what we have.

Some existing lanes could be given over to the goods-movement industry, which can also flourish with automation. Driver fatigue and rest rules would end. Safety would be guaranteed. Trucks could platoon—draft bumper to bumper—at high speeds in their dedicated lanes, cutting time and fuel. The robot Uber model may not replace ownership for freight and commercial vehicles, but all the other benefits of automation apply. For that matter, private ownership of robotic consumer cars would still continue for those who want it—and, at the outset, polls suggest many if not most Americans will not want to give up on a hundred years of pride, status, and habit linking them to car ownership. But the economics of a shared system and the practical benefits of having access to many different purpose-built cars instead of one jack-of-all-trades vehicle will be compelling. The Millennials will flock to it.

The upside for carmakers who seize this transformation as an opportunity instead of a threat are compelling, too: cars used around the clock instead of parked around the clock will wear out quickly and need to be replaced much more often than the ten-plus years Americans keep their privately owned cars today. Manufacturers of the new generation of purpose-built vehicles will replace the same car three or four times during those ten years. The first movers in this new competitive landscape will do very well by embracing this scenario. Today's carmakers may end up making that leap. Or newcomers will swoop in and steal the business from them, as computer makers replaced the typewriter manufacturers.

Driverless cars are not a total panacea. Enormous amounts of transportation would still be embedded in our daily lives and consumer economy. Much of the current pain would be vanquished, though.

The problem of traffic jams would disappear.

The problem of traffic carnage would be gone, deaths dropping from the tens of thousands range to something measured in the hundreds or less. No more drunks on the road. No more texting and wandering over the centerline. No more careening into telephone poles on dark rural highways.

The problems of capacity, of overload on freeways, of trucks backed up at the ports, of cars crawling painfully up the 405 and every other overburdened highway at rush hour, all go away, too. As does the need to expend fantastic amounts of money expanding capacity.

For mass transit systems already in place, the last-mile problem can be solved at last, while compelling new forms of automated mass transit could emerge.

The pollution, climate, and health costs that have long been attached to our cars—and that drivers have never been forced to pay—go away, too. We get to keep our cars—a different ownership model, perhaps, yet still cars—but the transition from polluting fossil fuels to clean electric vehicles finally makes practical and economic sense in the land of robot cars. Fossil fuel dependence, with all its environmental, economic, and national security implications, could be eased.

The death of parking would be particularly huge. Parking isn't just a drag on our time. It is a drag on the economy, a voracious land hog that grows with cancer-like relentlessness because most new real estate development includes a large, expensive, legally mandated parking component. Providing ample parking is a sensible-seeming policy but it comes with unintended consequences. For one, new parking in cities induces demand just like new freeway lanes, encouraging more people to drive into congested areas instead of making travel choices that reduce the need for parking (such as buses, subways, walking, or biking). Later, when tastes and trends inevitably shift and a once-thriving urban shopping center or other destination loses patrons, its underutilized or abandoned parking lot will remain in place for years, an oil-stained island of asphalt and wasted space all too familiar to American city dwellers. A survey of mixed-used districts in cities across the country found that, despite a perception of parking scarcity at individual venues, the number of spaces available in the immediate areas exceeds demand by an average of 65 percent.[3] Our oversupplied parking systems are yet another wildly inefficient use of resources in the door-to-door universe.

America's parking king, Los Angeles County, is littered with disused lots, which is no surprise in an area where a fantastic 14 percent of the incorporated land area consists of parking. That's the entire sprawling county of 87 suburbs ringing the City of Los Angeles, not just the concentrated urban core (where the percentage of land devoted to parking is closer to 30 percent). LA County boasts 18.6 million parking spaces in all. Nearly half that total consists of lots for business, industry, and government; 5.5 million spaces are taken up by off-street residential parking; and 3.6 million spaces can be found curbside.[4] That adds up to 3.3 spaces for every car registered in the county. The space devoted to cars no one is driving covers about 200 square miles, which is 1.4 times the amount of land set aside for LA County freeways and streets. If those spaces were put together into a single parking lot, it would cover an area large enough to hold four San Franciscos.

The age of driverless cars would return all that land to community use. Los Angeles would have four San Franciscos of space to play with. Such an amount of land would be—is—priceless.

The rise of the robots would finally solve the problem of our bankrupt highway trust fund—not through policy brilliance but out of sheer starvation. The national gas tax that already falls so far short of paying drivers' way would wither and die completely in a robot regime if, as many transportation scholars and researchers believe, the ascent of automation drives the rise of electric cars. No gasoline means no gas tax revenue at all, and Congress would be forced to find a new funding source. This would be the perfect opportunity to build user fees right into the emerging autonomous car sharing economy, just as airport funding is quietly built into every airline ticket today. It would be the drive more, pay more model we're supposed to have right now, this time done right. And if politicians need cover for imposing such fees, all they have to say is this shift to automation will end the scourge of car violence, the number one killer of our children. How do you put a price tag on that?

The hardest part will be creating a fair and equitable road tax system during the decades of transition from the old world of human-driven gasoline-powered cars to the new autonomous model. The greatest benefits accrue only after most of the cars on the road are robotic. Most of the costs imposed on the roads—health, safety, and the environment—are caused by the old-school cars we're driving now, which means, in a scrupulously fair world, their owners would pay a greater share of any new transportation tax than driverless electric car users during this transition. Polluters pay, that should be the rule. Safety risks pay. What could be more fair, more market-based than that? This would drive the shift to autonomy more quickly, but it would cause outrage, opposition, and real pain. Doing this part right and fairly would end up being much harder than building the technology itself.

Not everyone is entranced by this driverless car vision, to say the least. David A. Mindell, professor of engineering and the history of technology at the Massachusetts Institute of Technology, argues that the idea of full autonomy is impractical, even mythic, and that we would be better off striving for a perfect 50-50 blend of human and computer behind the wheel. He uses the 1969 Apollo moon landing as a prime example of the successful merging of human judgment and computer capabilities that led to a safe landing on the surface of the moon. Getting such a blending to work well and safely in everyday driving is a great technical challenge, Mindell argues, but the one most worth pursuing, because it keeps human judgment in the loop for those occasions when that enemy of robotic intelligence—the unexpected—arises.[5]

This is both a science-based argument and a seductive appeal to the human desire to retain control. But it doesn't jibe with practical experience to date, which has shown that the most hazardous moment for autonomous cars is the hand-off from robot to human control, particularly during some sort of emergency. Startle reflexes, the split-second delay that causes humans to "freeze" before shifting from inattention to concentration to reaction, could make a partially automated system the worst, not the best, of both worlds. "The hardest part is this transition when we have partial automation," Don Norman, director of the Design Lab at the University of California, San Diego, says. "This has been shown over and over and over again in aviation. People cannot keep their attention on what the task is. Because that is not the way we are built."

A second problem is Mindell's choice of examples: comparing the primitive Apollo moon-shot digital systems operated by the best trained space pilots in history to today's technology with an ordinary driver at the wheel. That is a bit like comparing that same Apollo command module to the Wright Brothers biplane. It's a mostly useless comparison, and not just because today's computers and software are light-years ahead of Apollo's or because

NASA pilots are trained to pay attention far more rigorously than ordinary car drivers or even because they were concentrating on the mind-blowing moment of *landing on the moon for the first time* versus a somewhat less historic drive to the local Denny's. No, the main problem with the comparison is that it ignores the critical role that detailed 3-D mapping of the driving environment plays in the success of the most advanced driverless car to date, the Google car. The moon landing was a first-time event. The Google car succeeds because it has "learned"—or been programmed—to safely navigate the most well-mapped terrain in the human universe, with over a million test miles under its digital belt, allowing it to adjust to the many variables that are added to that map by the presence of human drivers, cyclists, and pedestrians.

Even imperfect autonomy of this type will be far better than the distracted and imperfect human drivers roaming the highways today, Norman argues. "I am in favor of full automation. It's inevitable and when it comes will save huge numbers of lives, and it will turn out to be a blessing."

The Google car has mastered city driving. Grinding to a halt for a distracted computer-using jaywalker is flashy, but detecting and protecting a wayward pedestrian is actually a fairly easy challenge for the robot. Harder still is detecting a ball bouncing into the street, identifying it as a ball, then realizing that a child might follow that ball into the street and acting accordingly. The Google car can do that, too. But the most impressive achievement, wherein lies the true mastery of city driving, is how it handles the mundane but constantly varying minutiae of navigating through human traffic.

As I'm riding through the streets of Mountain View, just another car in the mix except for the cameras and radars and other distinctive moon-rover gear bolted to the Lexus test vehicle, the way the car handles a left turn blows me away. It approaches an intersection and signals a turn, then creeps forward without committing to the turn, just as a human does—not to get a better view, because this car sees everything, but to signal the intention to turn to other drivers. Then, the way clear, it executes the turn smoothly and moves on.

Exept for the fact that the driver is always attentive and never lead-footed, the experience is indistinguishable from driving with a human behind the wheel. And the reason for that is, in a way, a human *is* driving. The car doesn't have a supercomputer or advanced artificial intelligence tucked inside. What it does have are lines and lines of code through which human beings have instructed the car what to do at an intersection, when a jaywalker appears, when construction workers block a lane of traffic. That's why Google has driven its fleet of twenty-seven robotic test cars over a million highway and street miles throughout the Bay Area, and why they next moved on to Texas: to experience every possible situation and crisis on streets, so the car can be told by humans how to act. Google has even created what program director Chris Urmson calls his "red team"—a group of mischief makers who try to stump the car with unexpected obstacles or crazy behavior by other cars or cyclists who keep swerving out of their bike lanes. The robot just stops when it doesn't know what to do. But each new line of code eliminates one more stopper. This is how the robot "learns"; it's not about the hardware, which is fairly mature technology. It's all about the software, which is why a Silicon Valley company like Google can compete with Detroit carmakers, most of which have struggled with making software as good as their hardware.

There are still obstacles. Lidar is confused by reflections. The car doesn't know how to handle snow yet. Heavy rain is a problem, too. The spray of water that kicks up behind other vehicles creates a ghost image on the car's sensors that can appear to be a solid object, something to brake for. Medford says the coders have to figure out how to filter those false positives. There is, he admits, a long to-do list. First among them is the creation of detailed three-dimensional maps that the car uses to navigate. This is the other secret sauce of the Google car: its mapping prowess. These are the *Oxford English Dictionary* of maps, far more detailed than the Google Maps on smartphones and home computers. They measure curb height and sidewalk width, they note bike lanes and median dividers, and they take special note of distinctive markers in the landscape to aid in navigation with far greater precision than GPS coordinates. The maps are not hard to make, Medford says: "You just drive the car in manual mode and the computer builds the map." So far, the company has mapped all the freeways in the Bay Area and all the streets in the little city of Mountain View, population 77,846. All they have to do is map the rest of the country before bringing the car to market, Medford says. Most people might consider that daunting, but the folks at Google shrug it off as the least interesting challenge before them. "It's just driving," says Medford. "We'll get to it after we get the technology perfected."

There have been several accidents involving the Google car—none serious, and none, according to Google, the robot's fault. This is how the Google car's insistence on observing the speed limit and not running red lights can be a safety issue. It's been rear-ended several times, once because it was observing the speed limit on a street full of speeders, and once when it was properly stopped at an intersection when a distracted human plowed into it. Humans in both cars complained of neck pain afterward, but there were no serious injuries. The mix of autonomous and human-driven cars on the road may reduce accidents, but the greatest gains in safety likely will not occur until the vast majority of vehicles are driverless, Medford says.

The Google car has been pulled over by the police only once: for driving too slowly. It was traveling 25 miles per hour in a 35 mile per hour zone on a four-lane street in Mountain View (which is not technically illegal). The car was not malfunctioning: it is programmed to never exceed 25 miles per hour. The incident went viral, with the little Mountain View police department flooded with inquiries from media around the world. The car got off with a warning.

There's another wrinkle in this transition to vehicles smarter than the humans inside them: the idea of connected cars. This is a technology being pushed by the U.S. Department of Transportation, among others. It's different than driverless systems, but complementary to them. The idea is to leverage wireless technology to connect one vehicle to another—the digerati call this V2V—and to connect cars to the road and street infrastructure—V2I. Connecting cars would be a relatively simple matter, using inexpensive and proven transponders similar to the ones built into every airplane for decades, which transmit position, direction, and speed. These beacons could be anonymous if privacy concerns impede adoption, but the key would be how they could make an autonomous car aware of other vehicles that its sensors might not detect because of range, obstruction, or most relevantly, bad weather. This would be particularly valuable in a transitional mobility universe where autonomous cars and human drivers are mixing it up, as the robot cars would always know in advance if a human driver was slowing down to stop at an intersection—or on a course to just blow through.

The V2I concept would not be so easy to implement, because it would take many years and dollars to wire the built landscape and build beacons into our traffic signals, buildings, parking structures, and road signs. But such tech would solve the problem of sensor blindness in rain, fog, and snow. The road could literally talk to the cars, and let the autonomous driver know what's being obscured by adverse conditions.

There's no timetable or money at present for getting V2V or V2I up and running, and so all the driverless car projects that are being prepared for market now are designed to operate without any connection to anything—they are intended to be self-sufficient. This has the added bonus of greater digital security, as networked cars, like computers linked to the Internet, would be a far easier target to hack. But these same driverless cars will be able to jack into the connected car and infrastructure systems someday (if such systems ever become available), so security will be a continuing concern.

However it progresses, the transition to driverless cars will not be without some bumps and bruises, if only metaphorically. New technology and change always bring pain with the gains. For all its faults and harm, the hundred-year reign of the traditional car has had a powerful and positive impact on the economy and society, bringing unprecedented mobility and shaping our human landscape, culture, work, and lives for generations. This is part of the miracle of our door-to-door system of systems, and it is almost impossible to imagine replacing it with something else. Cars are linked in mind and deed with Americans' sense of personal freedom and opportunity.

But the current transportation regime is, literally, killing us—with the short-term bang and shatter of crashes, and the longer-term poison of pollution and carbon and oil dependence. Its inefficiencies cost us economically—more than we can pay for now, much less in the future.

The driverless future that could ease the killing and the costs, if it comes, will bring another kind of pain: lost jobs. Robots behind the wheels of cars and trucks will eliminate millions of human jobs and billions in salaries. Disruptive technologies always create new winners and losers, but it's not clear how or if the advent of automation will create enough new jobs to take the place of those lost. This is not new. Refrigerators put the ice man and his horse-drawn carriage out of business less than a century ago. Automatic washers put the laundry business down. Kodak's entire business used to be based on people shipping their cameras to the company to have the film inside developed and reloaded, then later customers just sent in the film. Now film photography is a near-dead business. Digital photography took a thriving analog technology with transportation deeply embedded within it and replaced it with a model that requires no transportation at all once the product is purchased. And now phones have eviscerated the stand-alone digital camera business as well.

The replacement of truck, bus, and cab drivers with automation will be wrenching, particularly since taxis have become an entry point into the workforce for immigrants, and truck and bus driving have provided one of the few enduring and plentiful blue-collar jobs that still provide reliable paths to middle-class prosperity. The American Trucking Associations reports that there are about 3 million truck drivers working in the U.S.—it's the single most common job in a majority of states—and about 1.7 million of that number are long-haul truckers, who would be most vulnerable to displacement by autonomous technology. The number of long-haul truckers is expected to grow by a minimum of 11 percent by 2022.[6] But the economic case for trucking fleets to go robotic is too powerful to ignore; this will happen, and likely far faster than consumer autos convert. This will be hard, but no more so than the fate of the once-robust web of smiths, farriers, horse dealers, feed stores, veterinarians, trainers, breeders, and stables that serviced every town and city in America until the 1920s, when human-driven cars displaced horses as the king of personal and commercial transport. Prior to that, during "peak horse," there was one working horse for every three Americans, and in New York City there were 297 horse-cart trips a year per person.[7] This transition was so recent that it occurred as the parents of the baby boom generation were growing up. Hundreds of thousands of jobs and thousands of businesses were wiped out when engines replaced horses as the backbone of human, freight, and farm transportation. This is what technological breakthroughs do. That's why they call it disruption. It's hard. It's upsetting.

But so are 35,000 deaths a year. So are 2.5 million trips to the ER a year. So are 5 million collisions a year. The single greatest cause of death for Americans ages one to thirty—that's disruption, too. That's pain. How much is ending that pain worth, particularly when the same solution also can take our roads and bridges out of overload and bankruptcy, toxins out of our air, and greenhouse gases out of the climate?

Still, Americans are uneasy about such change, and polls show most trust human drivers more than machines.[8] They may balk at such a transformation because cars, so long a part of the landscape and language of daily life, have become more than just transportation. We may hate the travails of transportation. We may hate the time we are stranded in traffic inside our cars. But we *love* our cars, that ultimate shipping container in which we ship ourselves and our families. We love them not as transportation tools but as objects and comforts and statements of style, the way a carpenter loves a well-made, well-used hand tool and is heartbroken when it's lost. It is the alchemy of cars we love, the interaction between hand and wheel and road, the gliding pleasure of a perfectly negotiated banked curve, the smooth and timeless sensation of having a world to yourself while driving at night, a sweet sound system swathing you in music and the solitude of movement. And, yes, for some—those who can afford it, or at least borrow to achieve it—there is the pleasure of having a trophy car that others may envy.

Many, perhaps most, American drivers will not want to give up these pleasures, these luxuries. But would they have to give them up in a world dominated by on-demand autonomous transportation appliances?

The short answer is no. First, any shift will be slow. The most optimistic predictions put 2030 as the soonest driverless cars are likely to become dominant. This would require firm action by government and industry to help make it happen. Given the inability of the government to add even a nickel to the gas tax to keep our bridges from collapsing, this seems unlikely. Trucking and delivery fleets may transition to autonomous driving that fast, because of the incredible economic benefits, and that may push passenger cars to make the leap more quickly. The year 2040 is a good bet for the rise of the on-demand robot passenger car. Whatever the time frame, with 265 million cars on the road in America, changing out that fleet would not and could not happen quickly or en masse. There will be a gradual transition.

Even then, just as people who find pleasure and fulfillment in riding horses haven't had to give that up, drivers could continue to enjoy taking the wheel. The evolution of horse travel provides the road map for the future of the human-driven car, the difference between transportation and recreation. Horses were once beasts of burden. They were our motors. Oats were big business, the horse era's equivalent of gasoline today (except the oats were a form of renewable energy). Now horseback riding is a luxury, enjoyed not on the regular streets and highways they gave up to cars and trucks, but in parks, on trails, and in equine-friendly communities. Cars will follow the same pattern. If there is a demand for it, there will be car parks and preserves and safe roads where humans can drive manually away from regular traffic. And if users want luxurious robot cars, the fleet owners will fill that demand, too. It will cost more, just as buying a regular luxury car costs more, but that's a model we're already used to. Far from denying drivers the joys they now find, the driverless scenario of the future could be the best of both worlds—cleaner, more efficient, less costly, and far less deadly.

It may be that automated vehicles' utility and opportunity will not be fully understood until they are out there in force. The smartphone experience may be instructive here. When Apple introduced the first iPhone, the design did not include open access to independent app developers. It was only when later iterations opened up the platform to outside innovators who thought up novel and new uses for smartphones that the true potential of such devices became clear. Autonomous cars are first and foremost a software product.

Google has designed with the car component maker Bosch and other partners a two-seater electric bubble of a car with no steering wheel or pedals, purpose built for short-distance city driving with a maximum speed of twenty-five miles an hour from its battery electric propulsion. It's cute and nonthreatening; the idea is to choose a city or two to test-market a robot car capable of transporting people in urban areas, and then go from there. Meanwhile, the big carmakers are focusing primarily on the highway automation feature; in the next few years, the future of door-to-door will be on display, and the next big change—as big as the first car or the first steam locomotive or the first airplane—will begin.

Google is a bit cagey about its own time frame. But it's not far off, says project leader Urmson, one of the small cadre of visionary engineers who have been driving robot car development since the first Defense Department competition in 2005 sent college teams into the desert with the first experiments in autonomous driving. Urmson has a simple goal: he says he doesn't want his son to have a driver's license.

His son turned eleven in 2015.

Chapter 13

THE NEXT DOOR

In the spring of 2015, the Museum of Contemporary Art, Los Angeles—MOCA, as most Angelenos call it—cohosted an event about the future of transportation with an organization known as Zócalo. The word is Mexican in origin, for "public square," and the partnership is intended to foster community conversations about the most vital topics of the day.

The invitation to this event was entitled: "Is Car Culture Dead?"

Ten years earlier, perhaps even five, this branding would have been more than merely provocative. It would have been ludicrous. Cars were the dominant species on the streets here, the T. rexes amid the field mice. Everyone knew that. Cars were the stars and, two decades after Missing Persons' sardonic lyrics first debuted, they seemed truer than ever: nobody walks in LA.

Yet, on this night, the auditorium was packed. In a city built for the car, where police night patrols study and sometimes question pedestrians because walking is, by definition, suspicious, people had flocked to discuss a world in which cars no longer dictated the transportation equations of daily life. At a museum. On a weeknight. Without a single celebrity in sight. The draw was a panel of eminent transportation scholars, activists, and car experts who had arrived in the following manner: one by bus, one by bicycle, one by walking, one by car but using a traffic app to map side streets free of congestion, and one by driving the freeway old-school—the only panelist to arrive late. "Traffic," he muttered.

Even the event moderator, Mike Floyd, editor in chief of *Automobile* magazine, who loves cars so much he based his career on extolling them, had sensed and even embraced a burgeoning transformation in transportation thinking. Not the death of car culture per se, he said, but a welling desire for more options than the four-wheel variety, and a sense that American cities were ready to welcome those options as never before.

"I do think the culture is shifting, very much so . . . a transition away from our car culture."

And yet, consider this juxtaposition. In this same city a few months before, the *Los Angeles Times* reported a disturbing increase since 2002 of hit-and-run collisions in which cars struck bicyclists and the drivers fled the scene. Car-on-bike hit-and-runs had gone up 42 percent between 2002 and 2012 (the last year for which data was available), leaving thirty-six people dead and 5,600 injured during that ten-year span—with almost no arrests or prosecutions.[1] The centerpiece of the report was the story of a man who was struck by a car while he was stopped at a red light in Beverly Hills. The driver sped off without lending aid, and the cyclist had been forced to drag his broken body to the curb, crying out for help. Eight of his vertebrae were fractured. His pelvis was shattered. Because of internal bleeding, he was kept in an induced coma for nearly a week. His medical bills exceeded $1 million, and it took him a year to recover, during which time he lost his job.

Two days after the crash, the driver showed up at the police department with a lawyer and photos of her car, admitting she had been involved in a collision with a cyclist, although she did not produce the car itself. The police declined to pursue the case without physical evidence from the car. Only the bicyclist's dogged efforts to persuade prosecutors to act, and the private investigator he hired to track down a witness, finally led to an arrest and prosecution a year and a half after the hit-and-run. After another year and a half, the driver pleaded no contest, received a sentence of 120 days in jail, served only two days—less time than the biker's coma—and was ordered to pay $638,000 in restitution. As of the date of the *Times* report, she had paid him $24.42 and had left the state. A lawsuit was working its way through the courts.

The response from readers was even more revealing than the original story. Letters poured in, expressing consistent and articulate outrage at—wait for it—bicyclists. "When it comes to cyclists involved in hit-and-runs," the paper's letters editor wrote, "readers didn't express much sympathy. Nearly all . . . expressed some empathy for the drivers."

Readers complained that cyclists don't stop for stop signs. They take up too much road space. They break the rules. They get in the way of the cars that actually belong there. If they were getting hit by cars, it most likely was the cyclists' fault, several letter writers suggested. No one expressed concern at the unenthusiastic response from the police and courts. The story of a cyclist's tragedy had instead become their opportunity to tee off on a topic they had been stewing about. One letter writer summed up the anti-bike zeitgeist this way: "I would guess that cyclists not playing by the rules of the road may upset the motorists and their emotions lead to some hit and run situations. Would it help if cyclists were required to take bicycle driving tests?"

Never mind that this story's main character was a cyclist stopped at a red light when a car slammed into him, or that official findings of fault in car-on-bike crashes run at about a fifty-fifty split, or that regardless of who is at fault in a collision, a hit-and-run is still a crime, not to mention an indecent and cowardly act. A man had to drag his broken body to the curb without anyone to help him, and all the letter writers could do was express anger at having to share the road with pesky cyclists, and suggest that such collisions could be avoided if bike riders would get over their sense of entitlement and be forced to pass a test. Because testing works so well in eliminating bad driving.

There you have it: one city, two very different visions of car culture; two radically opposed perceptions of what's wrong with our door-to-door system. Some crave change and more options and are hungry for the ability to use streets easily and safely without a car at least some of the time. Others abhor such change and deeply resent those who insist on what sounds like limits to their mobility, to the point that they are prepared to excuse hit-and-run drivers so long as the victim is riding a bike. One side argues that expanding bike lanes, plazas, and sidewalks to persuade people out of their cars will reduce overload simply by taking cars off the street. The other side views such notions as nonsense, for surely robbing cars of street space and giving slower bike riders and walkers travel equality can only increase congestion. How will drivers get to work on time with cyclists getting in the way? How will the police or ambulances get through during emergencies?

The same sort of divide exists between those who want to spend more transportation dollars on roads and those who favor more spending on public transit; between those who favor projects for passenger travel and those who want more spending to expand goods movement.

Where does a city—or neighborhood or nation—go from such seemingly intractable views of our door-to-door world? Is there a solution beyond hanging our hopes on the new and unproven wonder of autonomous cars and trucks, treading water while awaiting a technology that is unlikely to have a transformative impact before the year 2040? Can't something be done now?

The answer is yes. Of course. We've known how to fix most of the stuff that drives us crazy about driving—and biking and trucking and training—for decades. And many other countries have already done them to varying degrees. But if the question is *Will we do something now?* . . . well, that's another matter.

Over time, most of us have become acclimatized to the current sorry state of our door-to-door system, according to Janette Sadik-Khan, who swept into New York City in 2007 as the city's transportation czar, where she launched a whirlwind effort to persuade skeptical New Yorkers to embrace change that seemed, on the surface, to be a recipe for disaster. Her best argument was simple: *We've been doing the same thing to fix traffic for a half century. How's that working out for you?*

"For fifty years, everybody thought these are the ways our streets ought to be used. It's all about the asphalt. It's all about the cars. It's all about what happens behind the windshield . . . So initial skepticism is completely understandable. Because if your streets are frozen for fifty years, you accept them. If they're dangerous, if they're ugly, if there's no street life on them, okay: They're our streets."[2]

But Sadik-Khan, New York City's Transportation Department commissioner from 2007 to 2013, had the backing of the mayor who hired her. Billionaire businessman Michael Bloomberg gave her freedom to reinvent New York traffic. And she managed to do what many said was impossible: she redesigned iconic Broadway from Columbus Circle to Union Square, launched the nation's biggest bike share network, grafted four hundred miles of bike lanes onto some of the busiest streets in the world, put up red-light and speed cameras to try to catch the nation's most unruly drivers, built sixty pedestrian plazas across the city to revitalize street life, and, perhaps most spectacularly, closed Times Square to cars in order to restore pedestrian supremacy to some of the most valuable and recognizable square footage on the planet.

As the projects took shape, they were roundly condemned for creating a traffic Armageddon, Sadik-Khan recalls. "If you got your news from the tabloids, it was game over . . . And then the reality was, it actually made streets work better."

Like Carmageddon's transformation into Carmaheaven, the changes to New York's transportation bedrock defied expectations. After Times Square was closed to cars and converted to pedestrian plazas from Forty-Second to Forty-Seventh Street, many expected a traffic and safety disaster. Instead, she says, traffic on the surrounding streets either improved or remained the same, while accidents dropped 63 percent. Meanwhile, the business community's fears that retail sales would suffer proved to be unfounded: sales went up, and within two years 70 percent of retail managers supported the changes and wanted them made permanent.[3] It's common sense, really, as Sadik-Khan says. In a city, people on foot are customers. People in cars are just passing through.

The pattern was repeated throughout the city. Adding bike lanes and reengineering traffic flow for a calmer, steadier pace around Union Square improved traffic flow there 14 percent. A protected bike lane—not just a stripe on the road but a physically separated place for cyclists, the first in a major American city—led to a 50 percent reduction in congestion. The city has hard evidence to support this, using GPS data culled from the ubiquitous Yellow Cabs that ply the streets of Manhattan. The data showed, for example, that before a bike lane was placed on Columbus Avenue, it took four and a half minutes, on average, to drive from Ninety-Sixth Street to Seventy-Seventh Street. After the bike lanes came in, the transit time dropped to three minutes—with the same number of cars.

That sounds unbelievable, as it violates all the traditional rules of traffic. But Sadik-Khan and her new age traffic engineers always accompanied bike lanes with clever street design. Columbus Avenue was a one-way street with five lanes: three for traffic, one hybrid lane that was for traffic during rush hour and parking the rest of the time, and one for parking. Those lanes were all twelve feet wide, the standard width for 65-mile-per-hour freeway lanes. Lanes on city streets don't need to be so wide. By shrinking them to ten feet, enough room could be claimed for a six-foot protected bike lane and a five-foot buffer with landscaping between the parking lane and bike lane. The old configuration had drivers make left turns from traffic lanes but the new setup included pockets for left-turners so traffic didn't back up behind them. And so bike lanes don't speed up traffic by magic. They do so when combined with smart engineering. It just *looks* like magic.

Perhaps Sadik-Khan's most controversial move was the decision to install red-light cameras at 150 intersections citywide, plus speed cameras, some fixed, others roving, to catch (or deter) speeders and red-light runners. Decried as a plot to generate revenue rather than safety, similar systems are being dismantled as politically unpopular in communities nationwide. But Sadik-Khan says New York's data makes clear that the cameras are highly effective at curbing unsafe driving on city streets. At New York City intersections with red-light cameras, serious injuries dropped 56 percent and pedestrian injuries dropped 44 percent.[4] Areas near speeding cameras saw injury crashes drop by nearly 14 percent.

"It's absolutely a plot," says Sadik-Khan. "A plot to save lives."

When Houston removed its red-light cameras from intersections in 2010, crashes of all types at those intersections rose 116 percent, major crashes went up 84 percent, and fatal crashes rose 30 percent.[5]

"I can't think of a single other health issue where there is this much death and this much lack of concern," says Sadik-Khan. "Red-light cameras work. That's the plot."

New York's experience suggests that long-elusive, seemingly unattainable goals—safer and more prosperous streets, better traffic flow, and a (relatively) peaceful mix of walkers, cyclists, and cars, all moderate rather than massive infrastructure expenses—have been waiting there all along. In the past, traffic design and planning have been all about connecting two dots, giving politicians a sexy ribbon-cutting moment on a project that may or may not be worth the massive investment, and then flushing as many vehicles through as possible. Street designs have long been rated primarily by their LOS, their level of service, meaning that the more cars that can flow through a given intersection in the shortest amount of time, the better. LOS is why so many urban environments are indifferent, hostile, or unsafe for people on foot or bike. LOS gave us the stroad. And loss of LOS is why everyone expected Sadik-Khan to fail with her bike lane and pedestrian projects.

But after many years of pursuing LOS as traffic nirvana, the system is still chronically overloaded, dysfunctional, decaying, and nowhere near ready for the future. This was Sadik-Khan's pitch, and it was what allowed her to turn convention on its head to pursue her ideas and goals, furthered not through epic and costly projects as much as by shattering myths and using relatively simple off-the-shelf methods and technology to do better. Now other cities are adopting her template. Sadik-Khan started a movement.

In California, another rising star in the movement to reinvent traffic is Seleta Reynolds, whose job as general manager of the Los Angeles Department of Transportation is to do for LA what

Sadik-Khan did for New York—times ten. When LA Mayor Eric Garcetti hired her away from San Francisco in 2014, he branded Reynolds "the ideal field marshal in the war against traffic."

Hers is arguably the greatest transportation challenge of the era—at least people walk in Manhattan. Reynolds is charged with reengineering traffic to make walking as reasonable (and safe) a choice in LA, where some major streets have been denuded of pedestrians for decades, along with bicyclists, sidewalk diners, and casual transit riders. Her charge is to open up the streets to all of them, while also making it possible to drive LA without crippling traffic getting in the way. Oh, and one more little mission: While she's at it, Reynolds is charged with ending all traffic deaths in the city by 2025.

Still, Reynolds prefers to define her job less as field marshal and more as game changer, offering mobility alternatives after five decades of having only one practical option: the car. That's as much about perception as infrastructure, she says. Why is it, she wonders, that Angelenos can name their favorite restaurant, favorite museum, or favorite park in a heartbeat, but if you ask them their favorite street, they look at you like you're nuts? To its own people, LA streets are places of dread.

"It shouldn't be that way," she says, explaining what she believes is at the heart of her job. "Streets are spaces that belong to the people. They are public spaces, and we should be able to think about them like any of our other favorite places."

Her ideas—and other similar initiatives across the country—offer promise but no panacea. Even though bike lanes and pedestrian malls can improve safety, traffic, and mobility choices, they also lead to conflict that reveals just how intractable the fundamental problems of goods and people movement can be. In LA and New York City, the new bike-car-walking rules have unexpectedly pitted pedestrians against cyclists, who berate one another for failing to follow the rules of the road. New York's Central Park has become a particularly angry battleground after numerous high-profile collisions between bikes and walkers, including two crashes that left two pedestrians dead in 2014. The police have cracked down on cyclists by issuing hundreds of citations for rolling through red lights and failing to yield to pedestrians, but the bad feelings have yet to ebb (with the *New York Post* tabloid egging on the conflict with such headlines as: "New York's Cycles of Death: Our Arrogant-Biker Nightmare"). Other cities—San Francisco, for one—have grappled with aggressive cyclists who deliberately block cars and, in one videotaped confrontation in August 2015, surrounded and smashed a car after it bumped one of the blockade bikes.

The Times Square pedestrian mall, meanwhile, after drawing years of praise at home and emulation by other cities, has come under fire for attracting aggressive and costumed (and occasionally half-naked) panhandlers—an annoyance, perhaps, but tame in comparison to its past history of seediness and street crime. The new mayor and his police chief suggested in 2015 that the area be restored to its old car-centric design, setting off a political firestorm. The president of the Times Square Alliance business group fumed to the *New York Times*, "We can't govern, manage or police our public spaces so we should just tear them up. That's not a solution. It's a surrender."

The lesson here is that building a better door-to-door system is not just about the infrastructure. There are obstacles of habit and behavior to overcome as well: of tradition, belief, and the sense of entitlement that drivers, cyclists, and pedestrians all bring to the mix. The failure to acknowledge door-to-door behavior has led political and civic leaders to cling to costly "solutions" to traffic and transport that just aren't working, at least in their current forms.

Large investments in mass transit, such as the light rail and subway system Los Angeles has been constructing since the 1980s (primarily through voter-approved local taxes, not federal handouts), have had some success, but ultimately have failed to attract the hoped-for ridership (a common pattern nationwide). As of 2015, LA's Metro Rail system had exactly eighty-seven miles of track laid, with more under construction, at a cost of $12 billion over twenty-five years. About 7.2 percent of Los Angeles–area commuters use public transit, which is above the national average, but two-thirds of that is by bus, which means that $12 billion bought 2.4 percent of commuters for the new rail lines.

These sorts of investments have achieved nothing close to the pre–World War II levels of rail transit use, much less the mass devotion to streetcars and light rail of America early in the twentieth century, when cars were still in their infancy. A bit over 5 percent of Americans commute by mass transit of all types, which certainly has a measurable effect on traffic,[6] but a significant portion of that ridership comes from the 55 percent of commuters in the New York area who rely on transit. Recent mass transit investments nationwide have barely moved the needle on the 76 percent of workers who commute to work alone in their cars—suggesting the need to try something different, a rethinking of ways to make those investments pay off.

The same goes for the hundreds of billions spent to add lanes to highways without achieving the congestion reductions predicted and desired. This includes carpool lanes, such as the $1.3 billion for the Carmageddon addition in LA. Investment in carpool lanes continues despite the fact that proportionately fewer people than ever use them. In 1980, 19.7 percent of workers carpooled on their commutes. In 2013, the percentage of workers who carpooled was down to 9 percent.[7] This suggests we ought to be doing something different and more useful with the 3,000 miles of carpool lanes across America.[8]

Beyond the old strategies that aren't working, some trends tend to get ignored but bear examination. Consider the somewhat bizarre fact that Americans hate to walk more than any other people on earth. We quite literally walk less than everyone else, and far less than previous generations of Americans. This is not a minor thing but a trend that has profound implications for traffic planning and traffic jams (not to mention health and obesity). In 1980, 5.6 percent of Americans commuted to work on foot. By 2012, that number had dropped by half, down to 2.8 percent. Bicycling to work hadn't budged significantly in that same time frame: 0.5 percent in 1980 versus 0.6 percent in 2012.[9]

According to a 2010 study of American walking habits, in which 1,136 adults around the country wore pedometers for two days, the average American took 5,117 steps a day—about 2.5 miles. That's about half what a healthy person should cover every day, and about half the steps the average Australian or Swiss takes daily.[10] A separate study showed Amish men, who live as Americans did 150 years ago, take about 18,000 steps a day—nine miles. (Amish women cover about seven miles a day.)[11] Unsurprisingly, countries with higher average daily steps have lower adult obesity rates—anywhere between 3 and 16 percent of the population, compared to a 34 percent obesity rate in the U.S.

Will more walkable, welcoming, and safer streets change that? Or will getting people back on their feet require a greater cultural shift than just building it and hoping they'll come? It's a daunting question in Los Angeles, which should be a walker's and cyclist's paradise, with streets that are almost all flat and three hundred days a year with sunny skies and moderate temperatures. There is a small but hearty cadre in Los Angeles who choose to do what most Angelinos consider impossible, if not insane: They have no cars—by choice, not necessity. The demographic driving this trend is young professionals—Millennials, mostly—settling in the older sections of LA, the areas in and ringing downtown where streets and commerce are in the process or revitalization. With access to transit and more walkable streets than most parts of the city, these areas offer a rare opportunity to at least consider shedding the costs of car payments, parking, gas, and car insurance. That's what drew Monique Maravilla, a former stagehand for Cirque du Soleil and the LA Opera, who left stage work to open a coffee shop, Kindness & Mischief, on North Figueroa Street in the Highland Park section of LA. Now she bikes to work and most other places—a family lifestyle choice, as her fiancée has been car-free for a decade (though most of that time was in Oregon, she says, where such cities as Portland and Eugene have been more welcoming of car-free lifestyles).

What's it like getting around Los Angeles without a car? For Maravilla, it's a constant but mostly satisfying battle, like swimming against the tide. But she's been frustrated by her local city councilman's adamant opposition to installing bike lanes on busy Figueroa where her store is located, something he had promised to support in his election campaign. "It's too scary to bike on it now—the traffic is fast and drivers feel entitled," she complains. She had been counting on the added business that cyclists could bring to her fledgling shop. "Bikers stop and spend their money. The cars just zip by."

The health and economic benefits of car-free life keep her invested in the extra effort and planning it requires. Her friend Kacy Morgan, an LA sustainable transportation consultant and avid biker, has been car-light for two years for the same reasons, relying on mass transit and her bike most of the time, but renting a Zipcar or buying a Lyft ride when the only way to get somewhere is by car. "I save so much money compared to when

I owned my car," she says. "Hundreds a month. I don't ever see myself going back."

Still, this is not a common choice in Los Angeles, which is probably why Maravilla's councilman felt secure (or compelled) to flip-flop on promised bike lanes. Car owners are a much larger constituency, and many of them see bike lanes as traffic obstacles and competition. The long-term trends have been going in the drivers' direction for many years.

Half of all trips in the metropolitan area are under three miles—easy walking or biking distance—yet 84 percent of those short trips are made in cars.[12] A paltry 16,000 Los Angeles commuters get to work by bike. These trends are multigenerational, and we're passing them on to children. In 1969, just under half of American schoolchildren walked or biked to school. By 2009, the walkers and bikers had dropped to 13 percent.[13]

In all the many transportation plans made over the last fifty years, none saw this profound change coming, much less figured out what to do about it. Yet it is clear that the decline in walking and biking is a major contributor to traffic jams. All those kids who used to walk to school are now in cars—during rush hour.

Predicting transportation in the future is not, it turns out, one of the things America has been good at. When the interstate system was planned in the fifties and sixties, there were far fewer women in the workforce in America. So we're driving and moving goods today on a system that was not designed to accommodate one of the most disruptive changes to our transportation system of systems: today's workforce is split evenly between men and women. Instead of one household member driving off to work, it's now two—most often in separate cars. No one expected it when the backbones of today's transportation systems were built. No one imagined it. Planners simply assumed existing trends would continue.

The same holds true of the advent of e-commerce, of containerization, of burgeoning truck traffic pounding road surfaces much faster than anticipated, of a declining willingness to carpool or use transit or to even do such mundane daily trips as walking to school instead of driving. The nation simply didn't plan for any of it. We couldn't plan for it because our transportation crystal balls don't work very well, and so we just use current data and technology, add in population growth, and assume nothing else will change. Transportation planning, in short, is based on a fiction, an imagined steady-state universe. The result is the system we have today, part miracle, part madness, always teetering on the brink of overload.

"It's the 'Shut your eyes and stick your fingers in your ears and pretend nothing different will ever happen' method of planning," says David Levinson, the transportation scholar and a former planner himself. "It's kind of absurd, but we're still doing it that way."

In the summer of 2015, Los Angeles proved Levinson right when it published its official vision for the future of transportation in the second largest city in the country, an aspirational document called *Mobility Plan 2035*. It's an ambitious document that some city leaders have described as LA's new "constitution" of transportation, and others have vowed to oppose vigorously enough to keep the city's lawyers busy for a generation. But that's not the problem with the plan. The problem is that, for all its forward-thinking, paradigm-shifting aspirations, it was, like so many of our long-range transportation plans, obsolete the day it was released. This supposed blueprint of LA's door-to-door future twenty years down the line fails to consider, or even mention, the most disruptive and profound transportation scenario now on the horizon: driverless vehicles.

Los Angeles is not alone in planning tomorrow's transportation based on yesterday's technology. When the National League of Cities analyzed transportation plans for the nation's fifty most populous cities as well as the largest city in each state—a total of sixty-eight cities in all—only 6 percent had plans that considered the potential effect of driverless cars. Just 3 percent took into account the impact of such ride-sharing services as Uber and Lyft. Yet those services are up and running in sixty of the sixty-eight cities studied.[14] This is no simple bureaucratic omission. This is a disaster. These plans seek to determine how our cities will look, work, move, and grow in the future. Yet they ignore the most powerful and transformative trends in transportation in a century.

Perhaps that sort of lapse is inevitable when thinking of what the world should look like in twenty years. So much can—and will—change in two decades: twenty years before 2015, cell phones were dumb and heavy as bricks, and apps were the things you put on the table before dinner. Google didn't exist. The Web barely existed.

Twenty years before that, personal computers didn't exist, except as a kit you had to assemble. Gasoline cost 58.7 cents a gallon. Kids still walked to school.

Levinson says there's a simple solution: Make shorter-term plans. Set some ambitious but achievable goals that can happen in two to five years, and stop worrying about those big, bold, and mostly failed projects that look good at a ribbon cutting, then yield more bills than benefits. "Instead, choose the projects with the highest return on investment. You get a very different set of investments when you look for the one that has the most benefits instead of which ones are the sexiest or have the most political appeal."

So, what are these investments? It turns out they are not a secret. It's well-known what works. We could solve traffic. All you have to do is treat it like a market, subject it to the laws of supply and demand. All you have to do, in short, is price it correctly.

In other words, we need to revisit the worst four-letter word in the door-to-door business: the toll. Not for the revenue—though that wouldn't hurt, either, given the decline of the gas tax. It would be to change behavior.

Here's the problem. Peak times—primarily rush hour, when commuters go to work in the morning and come home in the evening—are when highways and streets are overloaded. People get stuck in traffic, lose time, waste gas, and have accidents. The movement of goods is affected and the economy suffers. The rest of the time, the off-peak drive times, most roads have plenty of capacity. Think of it as electricity: there's always plenty to go around except in hot summer days, when everyone turns on the air-conditioning and overload follows. Utilities charge extra for using more power at peak times—that's Supply and Demand 101—but the purpose of the higher rates is also to persuade people to reduce their usage to avoid overload.

Roads can work the same way. But right now there is no cost difference for driving at peak times rather than off-peak. Which is why more than half the trips during rush hour aren't work trips. People are going shopping or driving kids to school or going to breakfast (or happy hour) at peak times because they can do so without consequence. There is no mechanism to make them think about the delays they're imposing on the commuters and truckers who have no choice but to be there at that time. Putting tolls in place during rush hour, and taking them away at non-peak times, would provide an incentive for drivers to shift their optional non-work trips to different times.

"If just ten percent of those non-work-related rush hour drivers were persuaded to stay away during peak hours, congestion would disappear," Levinson says. "It's the single best thing we could do."

The technique is called congestion pricing, and it has worked in London, Milan, Singapore, and Stockholm. It's the cheapest cure for traffic jams there is, because it requires little new infrastructure and we know it works. It could even take the place of the desperately needed hike in the gas tax that Congress refuses to consider. But aside from a few small-scale demonstration projects, congestion pricing has not been politically viable in the U.S. Elected officials consider tolls to be toxic with voters, traffic cure or not, market based or not.[15]

The aversion to tolls extends to trucking, as well. A fully loaded heavy truck weighing in at 40,000 pounds does 10,000 times the damage to road surfaces than the average passenger car,[16] but in general they are not charged in tolls, fees, or gas taxes in proportion to their true cost to the infrastructure (the gas tax for heavy trucks is only a dime more a gallon than for cars). Tolling by weight would solve that inequity, but such proposals are routinely shot down as job killers that would raise the cost of goods. But in truth, giving trucks a pass on the damage they inflict to streets and highways is an enormous subsidy that the nation's taxpayers have to cover anyway.

Related simple solutions include persuading businesses to reduce rush hour travel with a simple time shift: have some employees start a bit earlier, others a bit later, staggering their commutes. Another way to tame rush hour would be for employees to work from home one or two days a week, even half days. Tax incentives for cooperative employers and tax penalties for uncooperative ones might be used. But there's no reason why employers couldn't do this simply out of a desire for happier, more productive employees who get to avoid driving in terrible traffic. This is a potentially zero-cost solution to congestion and, again, even just a 10 percent buy-in could lead to a permanent Carmaheaven. It is the ultimate low-hanging fruit of the transportation world:

time-shift commuters. It could even be as little as a half hour shift, because traffic behaves like a wave due to the habits of humans: rush hour traffic is heavier at the top of the hour than at the half hour.

Another cheap solution to traffic congestion would be the conversion of many—not all, but many—intersections from traffic lights to traffic circles. Mini-circles work well in residential areas instead of four-way stop signs. This is another simple strategy that, like tolls, met with an irrational hatred when proposed in American communities. But the circles—also called rotaries or roundabouts—are proven, with abundant evidence in the U.S. and abroad, to be far safer. They also speed up the flow of traffic by as much as 20 percent (a no-brainer: the cars don't have to stop and wait for lights to change). There are fewer crashes in circles, and those that do occur are at low speeds and never head-on. The kind of high-speed, blow-through-a-red-light, crushing impacts with other cars and pedestrians that happen daily at signaled intersections just don't happen in traffic circles. Circles are also cheaper: there are no lights to power or maintain, no sensors embedded in the streets, and nothing to fail during a blackout. There has been a slow uptick in the number of traffic circles in the country in recent years, but America is still the land of the traffic light.

Mass transit is another area that could be fixed far more cheaply than most believe. And the focus should be on the most humble of containers: the bus. Investing in bus transit gets the most bang for the buck, because the infrastructure—streets and roads— is already in place, which means the cost of adding service is far lower than trains and trolleys. But big rail projects are sexy, they draw in big federal grants, they're photogenic—but they're not solving traffic problems. They're not having a transformative effect on the door-to-door travails Americans face each day.

To understand what it would take to make the bus the new stars of transit, Levinson suggests we look to the streetcar, the former king of the transit world. Why did the streetcars that flourished before World War II fail? Why did LA abandon the biggest streetcar system in history? Simple: cars got in the way. As good as those old systems once were, people fled them because they became slower than cars. And that was because cars were allowed to drive on their tracks and get in the way. Then the trolleys had two causes of delay: stopping for passengers and stopping for cars. That made them slow. If trolleys had gone faster than cars, if the cars had been kept out of their way—if the tracks, which were there first, had been sacrosanct—the story might have ended differently. And communities that are now reviving trolleys are making the same mistake by allowing cars to drive on their tracks. It seems more convenient that way, says Levinson, but it's a recipe for failure.

Buses are in a similar position. They tangle with cars constantly, and end up losing. Buses always take longer because of the problem of stops plus traffic. No one who commutes by car will switch to bus riding because buses get them somewhere faster.

But what if they could? What if buses were given priority over cars—their own lanes on city streets? This could be justified on the grounds that one vehicle with fifty people onboard should have priority over a car with one person onboard. One bus can take fifty cars off the road. Decades ago, Los Angeles dreamed of streetcar lines or monorails running on all the freeways. Why not buses? Give them the carpool lanes. Let them go faster than the speed limit everywhere the freeways go. How many times would a commuter stuck in traffic have to see a bus zooming by, unimpeded, before he or she decided to give it a try?

This has been attempted in limited fashion in a number of cities with express transit ways, but why not make it the rule instead of the exception? If a bus is both faster and cheaper than a car, it would be a far more attractive option.

There's still the last-mile problem, and this is where the new dynamic of ridesharing services can complete the solution. The transit authority could partner with a rideshare service—or create its own—to get riders to bus stations (or rail stations, for that matter). Offer riders a package deal, a true door-to-door solution, at a rate that beats car ownership. It's possible. Lyft has launched a service called Lyft Line, a rideshare minibus that picks up multiple passengers and charges far less for the trip than for a single rider. Over half the Lyft rides in San Francisco are now Lyft Line rides. This is how public transit agencies need to think. This is how they can gain market share. Lyft executives are already publicly positioning the company as a partner with mass transit rather than a competitor. A rideshare–express bus combo will certainly cost a lot less than paying $12 billion for light rail with a 2.6 percent share of commuter ridership. If such a service could be fast, convenient, and affordable compared to owning a car and commuting alone at peak hours, it could change the door-to-door world in a big way and prepare the way for the driverless car (and driverless bus) of the future.

As for goods movement, where bus express lines don't make sense, give the carpool lanes to the trucks. Absent that, just relieving congested car traffic with tolls and bus ways would open up the road for the movement of goods. The reduction of congestion would reduce the need for new construction and free up funds to repair the existing roads and bridges that the goods-movement system desperately needs.

This is just a sampling of the sorts of high-impact, affordable solutions available. There are also the safety measures available now: collision avoidance systems that step in when drivers doze or are distracted or just make errors (systems that airliners have had for decades); automated breath analysis for preventing drunk driving; governors to prevent speeding. We could deploy them now if we had the will. In terms of the lives saved and medical costs avoided, they would pay for themselves, over and over. Driverless cars may make all this moot, but at best, that's a decade or two in the future. There are solutions available today.

We have indeed, reached a fork in the road.

Then there is the matter of global warming. It's impossible to understand the transportation embedded in our lives without acknowledging the carbon footprint attached to it and the meteoric increase in that footprint from the last forty years of globalization and off-shoring. That rise is a direct byproduct of our door-to-door economy—including the rise of China as the leading carbon polluter, where so many of the emissions are tied to the manufacture and transport of U.S. goods. This is not what anyone wants to hear. When it comes to global warming and transportation, the same strange psychology comes into play that enables drivers to shrug off a road death every fifteen minutes and a trip to the emergency room every 12.6 seconds. The inextricable link between our cars and climate change, like the endless parade of crashes on our roads, are simply perceived as the inevitable cost of doing business—our normal state, nothing to get excited about, nothing to do about it. Yet our fears over the disruption and cost that would arise from *halting* the carnage or the carbon are perceived all too acutely. Just discussing the possibility of scrapping our current way of moving ourselves and our stuff leaves Americans feeling either bereft, angry, or convinced that embracing change means giving up what we know and love best.

This fear of deviating from a "normal" that is literally killing us masks what should be obvious. As author Naomi Klein has so ably documented in her 2014 book, *This Changes Everything*, a shift to clean transportation and carbon-neutral energy through technology that's already available can create *more*, rather than fewer, opportunities for prosperity, security, and quality of life for the vast majority of American and world citizens. What would that require? Nothing less that this: a rapid shift to electrically powered and autonomous shared cars for shorter trips that aren't covered by walking or biking. Clean mass transit systems, built on some of the highway lanes now dedicated to conventional cars, would expand to cover a greater share of longer trips. Rail, trucking, and ships dedicated to goods movement could start reducing their carbon footprint by transitioning from bunker and diesel fuel to natural gas, then electricity and carbon-neutral biofuels as those sources ramp up. These moves would be powered by a revamped grid dominated by renewable power sources that are *already* price competitive with fossil fuels. Embedding more miles and energy in our products can no longer be the winning strategy. Removing miles, shrinking the distance or the carbon (or both) needed to go door to door, must once again be the low-cost, low-risk approach, as it has been for most of human history. This means an end to the silent but massive subsidy that excuses energy companies, carmakers, and consumers from paying the true price of fossil fuel dependence, including the costs it imposes on national security, the environment, the climate, healthcare, and human lives.

In the business world there would be big losers in this shift—the powerful fossil fuel industry. But there would be equally big winners in renewables, in producers of electric cars, autonomous vehicles, and electrical infrastructure. Tens of millions of jobs would be created to convert homes, ports, logistics centers, military bases, and factories to solar, wind, and bio-power with built-in energy storage for round-the-clock use and charging of our vehicles. Imagine offering America's 174,000 coal industry workers first dibs on those jobs, and everyone else a G.I. Bill–like free college or vocational education in exchange for signing up for five years of public service building this sustainable version of America, or exporting it abroad through a new green Peace Corps. We wouldn't be giving up our way of life. We'd be making it stronger. All we would be doing is a long-overdue upgrade of a filthy, creaking, overloaded, inefficient century-old model that has worked miracles but is now past its prime—so expensive that it's left us trillions of dollars behind in maintenance and repairs, even as it sabotages the delicate balance of our climate.

The door-to-door system created the modern consumer economy. Now it's breaking the world. Ironically, it's also in a position to help save that world with a cleaner, more efficient, and safer way to roll. We only have to make the correct turn at the fork in the road before us. And whichever way we end up turning, once again I suggest: Buckle up.

It may seem that the door-to-door system of systems is governed by forces outside the control of the consumer, the citizen, or the individual driver. But there are, in fact, ways in which an individual or a family can act meaningfully, because so much of the mobility equations of daily life come down to individual choice.

What can a person or a family do, then, to make a difference?

A good starting point is the amazing statistic that nearly half of all car trips in LA (and many other cities) are under three miles. Three miles make for a fifteen-minute bike ride, twenty at most. Or an hour walk at a reasonable pace. We could choose to not get in a car for some of those short trips to the store or the bank or the post office. We could get our kids walking and biking to school again and start talking to our city councils about making protected bikeways to our community schools, so kids can feel safe pedaling to class. Schools could give students physical education credit for biking instead of being driven to school. This thing against walking and biking is a cultural tic that we have the power to change. To see young men and women shoulder to shoulder with people in their fifties and sixties—50 percent of commuters—all of them biking in winter weather to work in Copenhagen, is quite a humbling experience. Copenhagen is a city rich with bike lanes and protected paths, and sheltered bike parking is everywhere. Bike travel is baked into the culture because the generation of adults in the 1970s made a commitment—to energy independence, to health, to a reduction in car deaths—and they forced themselves to get out of their cars and pedal. And their kids, and now their grandkids, are reaping the benefits. They just do it because they have always done it.

So it can be in your town. On your street. In your house. Just a few short trips at first, just to see. That's how it starts.

The new rideshare services offer another opportunity. The economics of car ownership—of insurance and depreciation and gasoline and parking—are such that, if you drive less than 10,000 miles in a year, you could save money by ditching your car and going rideshare for everything. If you live in a place where you might consider commuting by rail or bus if only the station were convenient to your home, you could use rideshare to get you to the transit stop, and the economic case grows stronger still for dumping your car. It would be great if the transit agency would offer this kind of service, but anyone could do it on their own. If your household is like the majority in America, with two or more cars, one of them probably fits the mileage profile to make rideshare the better deal. Save the money. Avoid the traffic jams. Avoid the stress. And to repeat: *Save the money.*

Many of our traffic problems stem from selfishness. The idiot who weaves in and out of traffic, making people brake to avoid his swerves on the freeway; who speeds up and tailgates, slams on the brake, then bulls into the next lane to pass—you see this person every day on the freeway. He makes traffic worse. He may get where he's going a bit faster, but everyone else is slowed down by his antics.

Don't be the selfish driver. Don't drive during peak hours if you don't have to. You'll avoid traffic jams and do your part to reduce congestion for everyone else. Do you really need a toll to tell you that? If you stick to non-peak driving, you'll burn less gas, suffer less wear and tear on your vehicle, and lower your risk of being in a crash. Not being selfish pays off.

We live as no other people have ever lived, with so much transportation embedded in our daily lives that it defies belief. If you have a broader concern about your consumption habits and the transportation footprint of products, that's not out of your control, either. The same strategies that people use to be less wasteful match up with reducing a personal transportation footprint: buying used things (cars, clothes, surfboards); shopping local; choosing reusable over recyclable and recyclable over disposable. Such choices could strip many miles out of your consumption. They add up. They exert influence.

I started walking more places a few years ago, when my family adopted our first rescue greyhound. I will not extoll the many virtues of these skinny speedy hounds here. Suffice it to say we have three now, and they need plenty of walking. As I am the family member who works at home, I'm the one they approach in the afternoon and stare at silently, all three of them ringing me with those pleading eyes, telling me it's walk time. At first this was a chore, an interruption. Then it became a habit and a source of joy. And now, when I am away for work or travel, I miss it. The three-mile walks we take twice a day are now a part of my daily life, a joyful part. We go to the park, or Main Street, or we just roam the neighborhood.

I call them my personal trainers, those greyhounds. There's one other benefit worth mentioning: I lost twenty pounds because those dogs got me to rediscover humanity's first method of moving door to door.

Yes, our brilliant, mad transportation system is at a fork in the road, the path to Carmageddon or Carmaheaven still up in the air. Transformative technologies and new methods are pitted against crushing congestion and the stubborn habits that sustain it. Solutions may well come from the top down, from Washington or Google or some other titanic enterprise. Or perhaps the real transformation will lie in the small choices we all make each day in how we consume and work and move door to door—choices that, like many small steps, add up.

Acknowledgments

I wish to thank everyone who accompanied me on this *Door to Door* journey, only a few of whom I can name here: Jay Isais of the Coffee Bean & Tea Leaf; Ken Kolosh of the National Safety Council; Don Fontana of Domino's and his whole Ontario crew; Geraldine Knatz of the University of Southern California and former head of the Port of Los Angeles; Elizabeth Warren of FuturePorts and the LOLs; Debbie Chavez, Reid Crispino, Steve Chesser, and Captain J. Kip Louttit of the Marine Exchange of Southern California; Bob Blair and Captain John Strong of Jacobson Pilot Service; Noel Massie, Karl Blackmun, and Don Chubbuck of UPS; Ted Trepanier of Inrix; Ron Medford of Google; Kevin McKnight of Alcoa; Christine Anne Melhart Slay of the Sustainability Consortium; and The Transportationist, David Levinson, of the University of Minnesota.

I am especially grateful to my editor at Harper, Hollis Heimbouch, who brainstormed this book with me from the start, and my incomparable literary agent, Susan Ginsburg of Writers House, who has had my back since my first book.

Finally, I must thank my beloved family, Donna, Gaby, and Eben. This is the first time our home has been mentioned in any of my books. I expect I'll never hear the end of it.

Appendix

This is a partial list of fatal crashes on February 13, 2015, compiled from media and police reports.

12:15 a.m., Milton, New Hampshire: head-on collision between SUV and station wagon kills three women and seriously injures a fourth. Cause: crossing centerline, possible drowsy or distracted driving.

2:30 a.m., Madison County, Tennessee: Seventeen-year-old is ejected from pickup truck after hitting tree on country road. No seat belt. Cause: road departure.

2:45 a.m., Lafayette, Louisiana: Eighteen-year-old college freshman dies after being struck by multiple cars on Interstate 10. Cause: pedestrian on freeway.

2:45 a.m., Collinsville, Oklahoma: Sedan crashes into fence and utility pole on country road, ejecting twenty-two-year-old driver. No seat belt. Cause: road departure, possible DUI.

3:45 a.m., Genesee County, New York: Car rear-ends snow plow on the New York State Thruway, killing thirty-three-year-old passenger. Driver suffers minor injuries and is arrested on multiple charges. Cause: suspected DUI, speeding, unlicensed driving.

4:00 a.m., Johnson County, Kansas: Thirty-three-year-old dies when he drives pickup truck into guardrail on Interstate 635 on-ramp, overcorrects, and crashes into a ditch. Cause: road departure.

4:00 a.m., Lansing, Michigan: Wrong-way driver on U.S. Route 127 dies after causing three separate crashes, and one other death. Cause: crossing centerline.

4:30 a.m., Greensboro, North Carolina: Thirty-two-year-old suffers fatal injuries when he drives off the road and into a wooded area where Interstate 40 splits. Cause: road departure.

4:45 a.m., Flagstaff, Arizona: Driver dies and passenger is injured when a semitruck hauling a load of beer runs into a culvert on Interstate 40. Cause: road departure.

5:30 a.m., Auburn Hills, Michigan: Reacting to erratic driving by another vehicle, a thirty-three-year-old driver loses control, strikes a median divider, and comes to a stop on Interstate 75. The driver is killed when a third car strikes the disabled vehicle. Cause: erratic driving.

6:15 a.m., Rapides Parish, Louisiana: A twenty-year-old dies when he rear-ends a semitruck. No seat belt. Cause: speeding, possible distracted driving.

6:18 a.m., Wichita, Kansas: A forty-six-year-old driver dies after she drives into a pillar while making a left turn. Cause: road departure, possible speeding.

6:43 a.m., Conroe, Texas: A seventeen-year-old high school student dies on State Highway 125 when his motorcycle collides with a car turning left into his path. Cause: failure to yield.

7:02 a.m., Harlingen, Texas: A forty-four-year-old man on a country road dies after he crashes his pickup truck into a concrete spillway. Cause: road departure.

7:45 a.m., Bridgewater Township, Michigan: One man dies and another is critically injured in a head-on collision between a sedan and a pickup truck. Cause: crossing centerline, possible distracted or drowsy driving.

8:00 a.m., Gladewater, Texas: A high-speed police chase triggers a series of crashes, an oil spill, and the death of the eighteen-year-old targeted by the pursuit. Cause: speeding, reckless driving, police pursuit.[1]

8:45 a.m., Taylorsville, Kentucky: An eighteen-month-old boy dies when his father backs his pickup truck over him. Cause: backover blind spot.

10:00 a.m., New Lennox, Illinois: A seventeen-year-old trying to cross the highway to his school is run over and killed after he steps into the path of a moving semitruck. Cause: pedestrian failure to yield.

10:14 a.m., Citronelle, Alabama: A twenty-five-year-old highway worker dies after a pickup truck runs into the construction trailer he was working in. Cause: driver error, possible distracted driving.

10:20 a.m., Jefferson County, Alabama: A forty-two-year-old

county transportation department worker is killed when he drives his dump truck onto a rail crossing and into the path of a freight train. Cause: failure to yield to train.

10:30 a.m., Ness City, Kansas: A tanker truck rear-ends a semitruck on State Highway 45 as the semi slows to make a right turn. The tanker driver dies on impact; the semi rolls over, injuring its driver. Cause: driver error, possible distracted driving.

11:00 a.m., Lancaster, California: A driver is ejected and killed after the left rear tire of his 1969 Land Cruiser comes off, causing the vehicle to depart the road and roll over. Cause: mechanical failure.

11:50 a.m., San Bernardino, California: A high-speed police chase leads to multiple injury collisions on Interstate 215, ending with police fatally shooting their twenty-three-year-old suspect. Cause: police pursuit for car theft.

Noon, Lakeland, Florida: The thirty-eight-year-old driver of a Lexus sedan dies when he veers across the grassy median on Interstate 4 and collides head-on with an oncoming semitruck. The truck driver is unharmed. Cause: crossing the centerline.

12:18 p.m., Ellisville, Mississippi: A two-year-old dies and his twenty-one-year-old mother suffers critical injury when their sedan is sideswiped by a dump truck as it changes lanes on Interstate 59, forcing the car off the road where it crashed into a tree. Cause: driver error, unsafe lane change.

1:10 p.m., State Highway 13, north of Manhattan, Kansas: A fifty-seven-year-old man dies when his sedan crosses into opposing lanes of traffic and collides head-on with a pickup truck. Cause: crossing centerline.

2:00 p.m., Kiski Township, Pennsylvania: An SUV makes a left turn in front of an oncoming car on Pennsylvania Route 56, killing the sixty-one-year-old driver of the other car and injuring a passenger in each vehicle. The SUV driver escapes injury and flees but is later arrested for multiple crimes, including homicide. Cause: DUI.

2:28 p.m., Mesquite, Texas: A fifty-six-year-old woman dies after driving her sedan out of her apartment complex, through a red light, and onto a busy avenue, where another car collides with her. Its two occupants suffer minor injuries. Cause: driver error.

2:30 p.m., Lackawannock Township, Pennsylvania: A car drifts into the grassy median of Interstate 80, rolls four times, and the forty-year-old woman driving is ejected and killed. No seat belt. Cause: crossing the centerline, possible distracted driving.

2:40 p.m., Northfield, Minnesota: An eighty-seven-year-old man driving a pickup truck dies, and his eighty-one-year-old wife suffers serious injury, when they collide head-on with a semitruck. The semi driver has moderate injuries. Cause: crossing the centerline.

2:47 p.m., Hamilton, Ohio: A thirty-four-year-old woman suffers fatal injuries when she drives her pickup truck through a highway intersection and into the path of an approaching dump truck. The broadside collision kills the woman, seriously injures her passenger, and forces another car off the road. Cause: driver error, failure to yield.

3:10 p.m., Lancaster County, Pennsylvania: A fifty-year-old woman dies after her SUV drifts into opposing traffic on the Route 462 bridge, colliding head-on with a compact car. The other driver suffers moderate injuries. Cause: crossing the centerline, possible distracted driving.

3:14 p.m., Tensas Parish, Louisiana: A pickup truck drifts off State Highway 568 at high speed, strikes a ditch and rolls. The nineteen-year-old driver is ejected and fatally injured. No seat belt. Cause: road departure.

3:20 p.m., Columbus County, North Carolina: A forty-six-year-old man is killed after he drifts onto the shoulder of State Highway 904, loses control, veers into the left lane and drives off the road, striking a tree. Cause: road departure, possible distracted driving.

3:50 p.m., Douglas County, Oregon: A fifty-one-year-old man is fatally injured after his speeding pickup truck drifts onto the grassy median of Interstate 5, causing the truck to roll and eject him from the driver's seat. Cause: crossing the centerline, speeding.

4:35 p.m., Fair Play, South Carolina: An eighty-seven-year-old woman suffered fatal injuries after she pulls her sedan out from a side road onto State Highway 11 and into the path of an SUV traveling northbound on the highway. The nineteen-year-old woman in the other car is uninjured. Cause: driver error, failure to yield.

4:40 p.m., Denver, Colorado: A seventy-eight-year old woman turning right at a busy intersection drives into a family in a crosswalk, killing a three-year-old boy and injuring his five-year-old sister and mother. The driver is charged with misdemeanor careless driving. Cause: driver error, failure to yield right of way to pedestrians.

5:00 p.m., Tucson, Arizona: A seventy-year-old woman is struck and killed while walking on the shoulder of Speedway Boulevard by a car traveling in the same direction. Cause: undetermined.

5:15 p.m., Portland, Indiana: A twenty-six-year-old woman dies while driving on the rural road where she lived when she drifts off into a ditch and is ejected when the car rolls. No seat belt. Cause: road departure, possible distracted driving.

5:30 p.m., Sequatchie County, Tennessee: A fifty-six-year-old motorcyclist suffers fatal injuries when he rear-ends a semitruck on State Route 111 near a scenic overlook. Cause: driver error, possible distracted driving.

6:00 p.m., Pike County, Georgia: A sixty-three-year-old man dies when he loses control on a curving road near his home, overcorrects, and rolls his pickup truck. He is partially ejected. Cause: driver error.

6:15 p.m., Desert Hot Springs, California: While walking across a busy avenue near his home, a forty-four-year-old is struck and killed by a passing car. Cause: poor street lighting, lack of crosswalk, area where cars routinely exceed speed limit.

7:15 p.m., Forsyth County, Georgia: An eighteen-year-old high school student is fatally injured when his pickup truck drifts off a country road; he overcorrects, then spins across the centerline, where he is struck broadside by a heavy-duty pickup truck.

Cause: road departure, suspected distracted driving due to cell phone use.

7:23 p.m., Kenneth City, Florida: A sixty-five-year-old man is run over and killed by an SUV while walking across a street near his home. The driver, who had three prior convictions for drunken driving, is arrested for DUI and manslaughter. Cause: DUI.

7:39 p.m., Village of Orange, Ohio: A fifty-two-year-old woman dies when her SUV drifts from the left lane to the right on Interstate 271, leading her to overcorrect, crash into the concrete center median, and roll over. She is ejected and dies of her injuries. No seat belt. Cause: driver error, probably distracted driver.

7:45 p.m., Donna, Texas: A fifty-year-old woman walking along a country road dies when she is struck by a passing SUV. Cause: undetermined.

8:30 p.m., Modesto, California: A forty-eight-year-old man walking along a busy four-lane boulevard is struck and killed when he tries to cross in the middle of the street, about thirty yards south of marked crosswalks and traffic lights. Cause: pedestrian error, possible intoxication.

8:30 p.m., New Braunfels, Texas: A sixty-five-year-old man suffers fatal injuries while jogging along a frontage road next to Interstate 35. The car and driver who ran him down flee, leaving him alive but unconscious, possibly for hours before he is found, too late to save him. Cause: hit-and-run.

8:45 p.m., St. Augustine, Florida: A fifty-nine-year-old man is killed after he appears to initiate a right turn but then veers left to make a U-turn on State Highway 16. The car behind broadsides the U-turning car. That driver is seriously injured. A passenger in each car suffers minor injuries. Cause: driver error, U-turn in traffic.

9:33 p.m., Marion County, Alabama: A thirty-six-year-old is killed when he drifts off the road and loses control of his pickup truck on Highway 253, causing the truck to roll, ejecting the driver. Two passengers are ejected and suffer serious injuries. No seat belts. Cause: road departure.

9:48 p.m., Long Beach, California: A fifty-five-year-old man in a wheelchair is fatally injured when a speeding luxury sedan crashes into him in a marked crosswalk, then flees. Cause: speeding, reckless driving, hit-and-run.

10:16 p.m., Council Bluffs, Iowa: A seventy-seven-year-old man who claims he blacked out drives a minivan into oncoming traffic on a busy avenue and crashes head-on into a sedan carrying a family of four. An eighty-four-year-old woman in the minivan dies, while five others in the vehicles were injured, two critically. Cause: crossing the centerline, incapacitated driver.

10:30 p.m., Chauvin, Louisiana: A twenty-eight-year-old man is walking home from a bar along the shoulder of State Road 58 when he is run over and killed. Cause: DUI.

10:30 p.m., West Palm Beach, Florida: A woman steps into the middle of a downtown street into traffic and is struck and killed by a passing hatchback. Cause: pedestrian failure to yield.

10:45 p.m., West Covina, California: A drunken driver speeding on a curving residential street in this Los Angeles suburb loses control and hits a curb, smashes into a tree, overturns, and slams into a retaining wall. The driver and a passenger in the front seat suffer only minor injuries, but a twenty-four-year-old backseat passenger is killed. No seat belt. Cause: speeding, DUI.

11:00 p.m., Winona, Missouri: A forty-five-year-old woman dies when she drives her compact car over the centerline on Route 19 and collides head-on with a pickup truck just five miles from her home. The pickup driver suffers serious injuries. Cause: crossing the centerline, probable distracted driving.

11:00 p.m., Reserve, Louisiana: A thirty-two-year-old man attempting to cross a state highway near a major intersection is struck and killed by a passing car. Cause: undetermined. Police rule out DUI and speeding.

11:20 p.m., Independence, Missouri: Two SUVs collide while both drive westbound on Interstate 70. The sixty-eight-year-old driver of one vehicle dies when his SUV leaves the roadway, overturns, and he is ejected. The SUV's occupants are uninjured. No seat belt. Cause: probable unsafe lane change or drift.

11:35 p.m., Wise County, Texas: An eighteen-year-old driver dies on a rural road when his car exceeds the speed limit, leaves the road at a sharp curve, crashes through a fence, rolls several times, then lands in a residential front yard, ejecting the driver. A passenger is seriously injured. Cause: speeding.

11:39 p.m., Lonoke, Arkansas: A married couple and their three-year-old daughter are killed when a heavy-duty SUV rear-ends their smaller sedan on Interstate 40, causing it to leave the road and crash into a tree. The SUV driver is injured. Cause: driver error, possible distracted driving.

11:40 p.m., Santa Ana, California: A dropped mattress in the fast lane of the 55 Freeway triggers a four-car chain reaction of crashes that in turn leads to a hit-and-run, with two people injured and one killed. Cause: object on freeway.

11:55 p.m., Reddick, Florida: A twenty-year-old dies when he drives his sedan around a curve and through a stop sign at the intersection of two country roads and continues off the road, crashing through a sign and a fence. Cause: speeding, possible distracted or drowsy driving.

Notes

INTRODUCTION: THE THREE-MILLION-MILE COMMUTE

1. Over the years, the stretch of I-405 closed for Carmageddon has been considered one of, and sometimes the worst, commute in America. The most current list from traffic data company Inrix, from March 2015, puts that section of I-405 at eighth worst in the country. Los Angeles has five of the top ten worst commutes in the country:
   1. Los Angeles: Riverside Freeway/CA-91 Eastbound
      Corridor: CA-55/Costa Mesa Freeway to McKinley Street
      Corridor Length: 19 miles
      Free-Flow Travel Time: 20 minutes
      Worst-Hour Travel Time (Friday, 4:00–5:00 p.m.): 81 minutes
   2. Chicago: I-90/I-94 Eastbound (Kennedy/Dan Ryan Expressways)
      Corridor: I-294/Tri-State Tollway to Ruble Street/Exit 52B
      Corridor Length: 15.9 miles
      Free-Flow Travel Time: 17 minutes
      Worst-Hour Travel Time (Friday, 5:00–6:00 p.m.): 72 minutes
   3. New York: I-95 Southbound (New England Thruway, Bruckner/Cross Bronx Expressways)
      Corridor: Conner Street/Exit 13 to Hudson Terrace
      Corridor Length: 11.3 miles
      Free-Flow Travel Time: 13 minutes
      Worst-Hour Travel Time (Friday, 4:00–5:00 p.m.): 63 minutes
   4. Los Angeles: I-5 Southbound (Santa Ana/Golden State Freeways)
      Corridor: East Caesar Chavez Avenue to Valley View Avenue
      Corridor Length: 17.5 miles
      Free-Flow Travel Time: 18 minutes
      Worst-Hour Travel Time (Friday, 5:00–6:00 p.m.): 63 minutes

5. Washington, D.C.: (I-95 Southbound)
Corridor: I-395 to Russell Road/Exit 148
Corridor Length: 23.9 miles
Free-Flow Travel Time: 23 minutes
Worst-Hour Travel Time (Friday 4:00–5:00 p.m.): 86 minutes
6. New York: (Long Island Expressway/I-495 Eastbound)
Corridor: Maurice Avenue/Exit 18 to Mineola Avenue/Willis Avenue/ Exit 37
Corridor Length: 16 miles
Free-Flow Travel Time: 16 minutes
Worst-Hour Travel Time (Friday, 4:00–5:00 p.m.): 53 minutes
7. Chicago: (Eisenhower Expressway/I-290 Eastbound)
Corridor: Maurice Avenue/Exit 18 to Mineola Avenue/Willis Avenue/ Exit 37
Corridor Length: 16 miles
Free-Flow Travel Time: 16 minutes
Worst-Hour Travel Time (Friday, 4:00–5:00 p.m.): 53 minutes
8. Los Angeles: (San Diego Freeway/I-405 Northbound)
Corridor: I-105/Imperial Highway to Getty Center Drive
Corridor Length: 13.1 miles
Free-Flow Travel Time: 13 minutes
Worst-Hour Travel Time (Friday, 4:00–5:00 p.m.): 53 minutes
9. Los Angeles: (Pomona Freeway/CA-60 Eastbound)
Corridor: Whittier Boulevard to Brea Canyon Road
Corridor Length: 21.7 miles
Free-Flow Travel Time: 22 minutes
Worst-Hour Travel Time (Friday, 5:00–6:00 p.m.): 61 minutes
10. Los Angeles: (Santa Monica Freeway/I-10 Eastbound)
Corridor: CA-1/Lincoln Boulevard/Exit 1B to Alameda Street
Corridor Length: 14.9 miles
Free-Flow Travel Time: 14 minutes
Worst-Hour Travel Time (Thursday, 6:00–7:00 p.m.): 49 minutes

2. The expansive, spare-no-expense drawing boards of the early sixties had called for two other north–south freeways to be built to relieve any bottlenecks on the 405. But those projects had been scrapped long ago as impolitic, impractical, or unaffordable.
3. Based on the 3.0156 trillion vehicle miles reported for 2014 by the Federal Highway Administration in *Traffic Trends*, December 2014.
4. This is a daily average calculated from the finding that U.S. freight move-

ment is greater than $20 trillion a year. From *Mapping Freight: The Highly Concentrated Nature of Goods Trade in the United States* by Adie Tomer and Joseph Kane, Brookings Institution, November 2014.

5. "Foreign Sources of Crude Oil Imports to California 2014," *California Energy Almanac.*
6. *Urban Mobility Scorecard Annual Report,* 2015, published jointly by the Texas A&M Transportation Institute and Inrix, August 2015.
7. *Injury Facts 2015 Edition,* the National Safety Council's annual statistical report on unintentional injuries and their characteristics and costs.
8. American Society of Civil Engineers' "Infrastructure Report Card, 2013."
9. The gasoline tax is 18.4 cents; the federal diesel fuel tax is 24.4 cents.
10. American Road & Transportation Builders Association.
11. Figures culled from "A Financial Model Comparing Car Ownership with Uber X (Los Angeles)" by Kyle Hill, Medium.com, August 31, 2014, and *Your Driving Costs: How Much Are You Really Paying to Drive?*, 2015 edition, American Automobile Association.
12. "On the Performance of the U.S. Transportation System: Caution Ahead" by Clifford Winston, *Journal of Economic Literature* (online), Brookings Institution, September 26, 2013.
13. "The Future Economic and Environmental Costs of Gridlock in 2030" by the Centre for Economics and Business Research, published by Inrix, July 2014.
14. According to traffic flow analysis by Inrix, as reported in "$1.1 Billion and Five Years Later, The 405 Congestion Relief Project Is a Fail," *Los Angeles Weekly,* March 4, 2015.
15. University of Michigan Transportation Research Institute.
16. "Uber Won't Kill Car Sales, but Ride-Sharing May Affect What We Buy" by Zach Doell, Kelley Blue Book, July 28, 2014.

CHAPTER 1: MORNING ALARM

1. Just to be clear, this is not an homage to phones made by Apple or any other brand; the same tasks could be accomplished with a variety of other smartphones (or tablets or computers). This just happens to be the phone I'm using at the time of this writing and, more to the point, Apple Inc. began in 2012 to make public considerable details on its major suppliers as part of a corporate transparency and sustainability initiative. These disclosures make it possible to piece together a large part of the iPhone's journey from raw material to complete product. That was the same year Apple was

singled out for criticism over poor working conditions at its top Chinese supplier, Foxconn (which happens to be a top supplier for many of Apple's competitors, too). The supplier information is on the Apple website, https://www.apple.com/supplier-responsibility/our-suppliers/site.

2. "Carbon Footprint of a Single Newspaper Equals One Kilometer in a Car," *Pulp & Paper Canada*, March 29, 2011.
3. From *Consumer and Mobile Financial Services 2013*, Board of Governors of the Federal Reserve, March 2013, and *2013 Fiserv Consumer Trends Survey Shows Growth in Mobile, Tablet Banking*, Fiserve, Inc., June 2014.
4. As of 2013, per Kidscount.org.
5. *6 Facts About Americans and Their Smartphones*, Pew Research Fact Tank, April 1, 2015.
6. "Is 30 Percent of Traffic Actually Searching for Parking?" by Paul Barter, Reinventing Parking, October 7, 2013, http://www.reinventingparking.org/2013/10/is-30-of-traffic-actually-searching-for.html.
7. The information on the assembly of the home button components is from Apple's Supplier Responsibility Report, https://www.apple.com/supplier-responsibility/accountability/.
8. Allegations of poor worker treatment at Foxconn gained high media exposure through a 2012 stage play entitled "The Agony and the Ecstasy of Steve Jobs," which was presented as nonfiction and reported as such by numerous media outlets, including *The New York Times* and *This American Life*. Both media organizations later branded the allegations as fabrications. (See Episode 454, "Mr. Daisey and the Apple Factory," *This American Life* (online), January 6, 2012; and "Mike Daisey Apologizes for Falsehoods in Monologue About Apple," *The New York Times,* March 26, 2012.)

   A subsequent audit requested by Apple and performed by the Fair Labor Association found overtime and wage violations were routine at the massive factory complex, which Apple and Foxconn vowed to remedy. (See "Independent Investigation of Apple Supplier, Foxconn: Report Highlights," Fair Labor Association, March 2012, via www.fairlabor.org.)
9. Other components include: a barometric sensor and accelerometer from Germany; the Corning "Gorilla Glass" from Kentucky; the five different power amplifiers from California, Massachusetts, Colorado, North Carolina, and Pennsylvania manufacturers; the motion processors from Silicon Valley; the Near Field Communications controller chip from the Netherlands; signal controllers and microphone parts from Ireland; the Sony cameras, LCD display, random access memory, antenna switch, and

touch-screen components from five separate suppliers in Japan; the antenna tuner from North Carolina; the flash storage chips from Korea; the audio chips and touch transmitter from Texas; the touch-screen films from 3M plants in the U.S.; the e-compass from Taiwan; the Simplo battery from Jiangsu, China; and a variety of other connectors, controllers, and modules sourced from cities across China and Taiwan.

10. Rare earth minerals in the iPhone: rareelementresources.com.
11. "Mining Your iPhone: Recycling iPhones Yields Gold, Silver, Platinum, and More" by John Koetsier, *Venturebeat*, April 3, 2013.
12. "The Humble Hero," *Economist*, May 18, 2013.
13. The most authoritative account of the development and impact of the modern shipping container is *The Box: How the Shipping Container Made the World Smaller and the World Economy Bigger* by Marc Levinson, Princeton University Press, 2006.
14. *International Shipping Facts and Figures*, Maritime Knowledge Center of the International Maritime Organization, March 6, 2012.
15. "Shipping and Climate Change: Where Are We and Which Way Forward?" *Policy Brief*, International Transport Forum, October 2015 (revised).
16. "iPhone 6 Plus Environmental Report," Apple Inc., September 2014.
17. Dairy industry white paper, Blu Skye Sustainability.
18. California Sustainable Wine Growing Alliance.
19. "Six Products, Six Carbon Footprints," *Wall Street Journal*, October 6, 2008.
20. "Estimated U.S. Energy Use in 2014," Lawrence Livermore National Laboratory, flow chart published online in 2015.
21. "Total U.S. Greenhouse Gas Emissions by Economic Sector in 2013," U.S. Environmental Protection Agency (online).
22. "Estimated U.S. Energy Use in 2014." Livermore's data shows that the transportation sector consumes 27.1 quads of energy a year (out of total national consumption of 98.3 quads). Of that, 21.4 quads are "rejected," i.e. wasted—78.9 percent of the total transportation energy picture. One quad is a fantastic amount of energy, equal to 10^{15} BTUs, or 252 megatons of TNT. By comparison, the most powerful nuclear bomb in the U.S. arsenal in 2015 is 1.2 megatons.
23. Industry data from the American Trucking Association.
24. See "The Boomerang Effect," *Economist*, April 19, 2012, and "Why Apple Will Never Bring Manufacturing Jobs Back to the U.S." by David Goldman, Money.CNN.com, October 17, 2012.

25. *Capturing Value in Global Networks: Apple's iPad and iPhone* by Kenneth L. Kraemer, Greg Linden, and Jason Dedrick, University of California, Irvine; University of California, Berkeley; and Syracuse University.
26. "Why Apple Will Never Bring Manufacturing Jobs Back to the U.S." by David Goldman, Money.CNN.com, October 17, 2012.
27. "Professional Trends: Supply Chain Management Schools," educationnews .org, http://www.educationnews.org/career-index/supply-chain-management -schools/.

CHAPTER 2: THE GHOST IN THE CAN

1. "Facts at a Glance—2013," *Industry Statistics*, Aluminum Association, December 2014, and "Aluminum Can Infographics Gallery" at aluminum.org.
2. Based on 2015 production of 53 million tons of primary aluminum and a market price of 77 cents/pound.
3. If storage were the only consideration, the cheapest, strongest shape for a single-serve beverage container would be a sphere, which would use the least material and distribute weight and pressure evenly throughout the inner surface of the can, avoiding the engineering weak points of corners and seams. But they would be a nightmare to pack in a case or carton: stacked, 26 percent of the space inside the shipping container would be dead space. (They'd also be clumsy to handle for consumers, and would be prone to rolling off tables.)
4. *Study Finds Aluminum Cans the Sustainable Package of Choice*, Aluminum Association, May 20, 2015.
5. From a sustainability perspective, aluminum cans are still not "green," as they are still single-use containers for which energy and transportation must be expended to recycle. A less wasteful option is to reuse containers.
6. According to the Aluminum Association.
7. According the Aluminum Association, aluminum can scrap as of February 2015 was worth $1,491 per ton on average versus $385 per ton for the plastic most commonly used for beverages (PET) and the zero dollars per ton recyclers are willing to pay for glass.
8. Federal mandates call for light-duty vehicles—passenger cars and pickups—to achieve average fuel efficiency rates of 54 miles per gallon by the year 2025. As of the 2013 model year, the average had reached 24.1 miles per gallon. "Light-Duty Automotive Technology, Carbon Dioxide Emissions, and Fuel Economy Trends: 1975–2014," U.S. Environmental Protection Agency, http://www.epa.gov/oms/fetrends.htm.

9. According to the Steel Recycling Institute.
10. "Steel vs. Aluminum: Which Wins for Fuel Efficiency AND Cost?" by Stephen Edelstein, GreenCarReports.com, October 13, 2014.
11. Red mud is one of the biggest waste problems facing the mining industry, with 80 million tons of it created annually, which must be kept segregated in giant holding ponds. If let loose, red mud can render lifeless any natural environment it touches—as demonstrated by a fatal spill in Hungary that swept through farms and a village, killing 10 and injuring 120 people in 2010. Sources: "Reducing the Environmental Impact of the Aluminum Industry" by Dr. Tim Johnson, Tetronics.com, and "Toxic Red Sludge Spill from Hungarian Aluminum Plant 'An Ecological Disaster'" by David Gura, National Public Radio, October 5, 2010.
12. "Review of Maritime Transport: 2014," United Nations Conference on Trade and Development, annual report.
13. Hall's patents prevailed in the U.S. while Héroult's patents were recognized in all other countries.
14. Oxygen is by far the most common element, not because of its presence as a life-giving gas in the atmosphere, but because so much more of it is chemically locked inside rocks and water, representing nearly half the earth's mass. Silicon is the next most abundant element, constituting 28 percent of the planet, and although we think of it these days as the expensive stuff of computer chips and solar cells, silicon is more commonly a principal ingredient in rocks and sand. Third-place aluminum weighs in at 8 percent of the earth's mass and is contained in nearly two hundred different kinds of rocks, common and precious alike, including sapphires, rubies, and emeralds.

    There are only a few other common elements: iron makes up 5 percent of the earth's mass; calcium, 4 percent; potassium and sodium, 3 percent each; and magnesium, 2 percent. The ninety other naturally occurring elements combined make up less than 1 percent of the earth's mass.
15. Powders and salts *containing* aluminum have been known and used for medicine and commerce for thousands of years, although no one suspected the metallic element at their heart. The ancient Greeks used one aluminum compound to sterilize and cleanse wounds; another, aluminum potassium sulfate—commonly called alum—was used for 5,000 years as a mordant to fix dyes to fabric. English wool merchants in the Middle Ages went to great lengths and expense to import the fine white powder from the Middle East because dyed wool fetched a far better price than plain wool. But the metallic element in this powder remained hid-

den. This is why aluminum is sometimes referred to as the most "modern" of metals—which is to say it's the metal humanity only recently figured out how to exploit, as opposed to the ancient forging of copper, tin, silver, and iron mastered thousands of years ago, primarily to make better weapons of war.

16. "Primary Aluminum Smelting Power Consumption," World Aluminum Organization, July 31, 2015, http://www.world-aluminium.org/.
17. "Steel vs. Aluminum: The Lightweight Wars Heat Up" by Jim Motavalli, CarTalk.com, February 3, 2014.
18. *Life Cycle Assessment—Energy and CO_2 Emissions of Aluminum-Intensive Vehicles* by Sujit Das, Oak Ridge National Laboratories, January 15, 2014.
19. Aluminum Association.
20. Alcoa.
21. U.S.-based Ball announced in 2015 its intention to purchase UK-based Rexam. The proposed merger was under scrutiny by European antitrust authorities.
22. *The Aluminum Can Advantage: Key Sustainability Indicators 2015*, published May 2015 by the Container Manufacturing Institute and the Aluminum Association. The report puts America's consumer recycling rate of aluminum cans at 66.7 percent in 2014.
23. "Me, My Car, My Life . . . in the Ultraconnected Age," *KPMG Reports*, November 2014, and "Study: Nearly 35 Percent of US Households with a Vehicle Have at Least Three," GreenCarCongress.com, February 12, 2008.

CHAPTER 3: MORNING BREW

1. Americans drink more soda, bottled water, and beer by volume, but a higher proportion of American adults—nearly two-thirds—say they drink coffee every day more than any of these others beverages. Source: Gallup Polls.
2. "Coffee Grinds Fuel for the Nation," *USA Today*, April 9, 2013.
3. WorldsTopExports.com.
4. International Coffee Organization.
5. *Top 10 Commodities*, data from 2012, Food and Agriculture Organization of the United Nations, Statistics Division.
6. The Sustainability Consortium, Christy Anne Melhart Slay, commodity mapping project.
7. Based on 2014 data from *The Current State of the Global Coffee Trade*, International Coffee Organization, August 20, 2015.

8. Assuming two coffee drinkers in the household making one pot a day in a typical electric drip coffeemaker, they would use about a pound of coffee a week.
9. "Origin and Genetic Diversity of *Coffea arabica* L. Based on DNA Molecular Markers" by P. Lashermes, M. C. Combes, J. Cros, P. Trouslot, F. Anthony, A Charrier, Proceedings of the 16th ASIC Colloquium (Kyoto), 1995.
10. *The World of Caffeine: The Science of and Culture of the World's Most Popular Drug* by Bennett Alan Weinberg and Bonnie K. Bealer, Routledge Books, London, 2001.
11. Data provided by Christy Anne Melhart Slay and Philip Glasgow Curtis of the Commodity Mapping Project at the Sustainability Consortium, headquartered at the University of Arkansas.

CHAPTER 4: FOUR AIRLINERS A WEEK

1. "Cool Climate Network, Renewable & Appropriate Energy Laboratory," University of California, Berkeley, http://coolclimate.berkeley.edu/calculator.
2. *U.S. Smartphone Use in 2015* by Aaron Smith, Pew Research Center, April 1, 2015.
3. *Record 10.8 Billion Trips Taken on U.S. Public Transportation in 2014*, American Public Transportation Association, March 3, 2015.
4. "Annual Sales of Light Duty Vehicles Are Expected to Exceed 122 Million by 2035," Navigant Research, July 2014 press release.
5. According to the EPA, the average vehicle weight for model year 2012 was 3,977 pounds, a 150-pound decrease from the model year 2011.
6. Hybrids remain a tiny minority of the U.S. fleet, less than 3 percent, and all-electrics are but a rounding error. U.S. Energy Information Administration.
7. Culinary Institute of America professor John Nihoff, quoted in "Car Cuisine," by Jaclyn Schiff, Associated Press report at CBSNews.com, November 9, 2005.
8. Eran Ben-Joseph, professor of landscape architecture and urban design at MIT, and author of *Rethinking a Lot: The Design and Culture of Parking*, MIT Press, 2012.
9. The City of Los Angeles estimates that 40 percent of city sidewalks are in disrepair, and Mayor Eric Garcetti announced early in his tenure that fixing them would be a priority.

10. *A New Direction: Our Changing Relationship with Driving and the Implications for America's Future,* U.S. PIRG Education Fund, Spring 2013. The report cites these factors in support of the Millennial Generation's move away from the car:
    - The Millennials (people born between 1983 and 2000) are now the largest generation in the United States. By 2030, Millennials will be far and away the largest group in the peak driving age thirty-five- to fifty-four-year-old demographic, and will continue as such through 2040.
    - Young people age sixteen to thirty-four drove 23 percent fewer miles on average in 2009 than they did in 2001—a greater decline in driving than any other age group. The severe economic recession was likely responsible for some of the decline, but not all.
    - Millennials are more likely to want to live in urban and walkable neighborhoods and are more open to non-driving forms of transportation than older Americans. They are also the first generation to fully embrace mobile Internet-connected technologies, which are rapidly spawning new transportation options and shifting the way young Americans relate to one another, creating new avenues for living connected, vibrant lives that are less reliant on driving.
    - If the Millennial-led decline in per capita driving continues for another dozen years, even at half the annual rate of the 2001–2009 period, total vehicle travel in the United States could remain well below its 2007 peak through at least 2040—despite a 21 percent increase in population. If Millennials retain their current propensity to drive less as they age, and future generations follow, driving could increase by only 7 percent by 2040. If, unexpectedly, Millennials were to revert to the driving patterns of previous generations, total driving could grow by as much as 24 percent by 2040.
11. A conventional internal combustion car loses most of the energy produced by burning gasoline through waste heat and mechanical inefficiency, using about 20 percent to move the vehicle. The inefficiency is compounded because that remaining 20 percent is expended on moving 4,000 pounds of metal in addition to the human driver. So only about 1 percent of a car's gasoline cost is actually involved in moving the person inside. (By way of comparison, a modern all-electric vehicle wastes only about 10 percent of its energy, using about 90 percent to move the car. Electric motors are vastly more efficient.)
12. "Air pollution and early deaths in the United States," Fabio Caiazzo, Akshay Ashok, Ian A. Waitz, Steve H.L. Yim, Steven R.H. Barrett, MIT,

Atmospheric Environment, Volume 79, November 2013. The study found a total of 200,000 premature deaths annually in the U.S. from combustion emissions, with the most death attributed to transportation emissions. Second, with 52,000 premature deaths, were emissions from power plants. Industrial emissions accounted for 41,000.

13. *Shared Autonomy* by Adam Jonas, Morgan Stanley Research, April 7, 2015.
14. "A Financial Model Comparing Car Ownership with Uber X (Los Angeles)" by Kyle Hill, Medium.com, August 31, 2014, and *Your Driving Costs: How Much Are You Really Paying to Drive?*, 2015 edition, American Automobile Association.
15. The U.S. Environmental Protection Agency calculates that 27 percent of annual carbon emissions in America are from transportation. Only the generation of electricity accounts for more: 31 percent of the total. In reality, transportation emissions could easily be the greater, as these figures specifically exclude the massive carbon emissions from cargo ships that bring us 90 percent of our consumer goods.

    According to the Lawrence Livermore National Laboratory, 80 percent of the energy expended for transportation is wasted. This means most of the oil burned in the U.S. inside cars and trucks achieves nothing in terms of movement, but contributes considerably to the creation of smog, lung disease, global warming, consumer costs, dependence on foreign oil, and profits for the oil industry.
16. Cars and light trucks (mainly pickups) make up 60 percent of U.S. transportation-related carbon emissions. Medium and heavy trucks contribute 23 percent. Rail and ships contribute 2 percent each, and everything else that moves and burns fossil fuels takes up the remnant. (This total for ship emissions is misleading, too, for it includes few cargo ship emissions, which occur in international waters and are generated by non-US sources. Even though the shipping is on behalf of the U.S., bringing goods to Americans, the emissions aren't counted in our tally or any other countries. Those emissions, immense as they are, don't exist when it comes to the accountants' view of global warming.) From: "U.S. Transportation Sector Greenhouse Gas Emissions, 1990-2013," U.S. Environmental Protection Agency, October 2015.
17. "Making driving less energy intensive than flying," Michael Sivak, University of Michigan Transportation Research Institute, January 2014.
18. As of October 2015.
19. National Safety Council data.

20. Air and vehicle trip data is from the U.S. Department of Transportation's Bureau of Transportation Statistics.
21. Centers for Disease Control, *Leading Causes of Death* reports.
22. National Safety Council, *Odds of Dying*.
23. *The Economic and Societal Impact of Motor Vehicle Crashes, 2010* (revised), National Highway Traffic Safety Administration, May 2015.
24. "Fact Sheet: America's Wars," U.S. Department of Veterans Affairs, November 2014, http://www.va.gov/opa/publications/factsheets/fs_americas_wars.pdf.
25. Centers for Disease Control, based on 2012 accident data, released October 7, 2014.
26. National Safety Council.
27. "Fact Sheet: America's Wars," Department of Veterans Affairs.

CHAPTER 5: FRIDAY THE THIRTEENTH

1. *Public Road and Street Mileage in the United States by Type of Surface*, Bureau of Transportation Statistics, U.S. Department of Transportation. Figures are for 2013: Of the 4,071,000 miles of roads, 2,678,000 are paved and 1,394,000 are unpaved.
2. These numbers are averages, based on most recent data from the National Safety Council. In reality, some days of the week and hours of the day have higher accident rates, while others have lower.
3. Data sources: National Safety Council for fatality and crash statistics; Centers for Disease Control for emergency room visits.
4. The seminal study of crash causation was the *Tri-Level Study of the Causes of Traffic Accidents* published in 1979 by Indiana University's Institute for Research in Public Safety, which found human error a definite or probable cause in 90 to 93 percent of crashes. The UK's Transport and Road Research Laboratory put the figure at 95 percent in a 1980 study. *The Relative Frequency of Unsafe Driving Acts in Serious Traffic Crashes*, the National Highway Traffic Safety Administration's study published in December 1999, concluded that in 99 percent of crashes studied, "a driver behavioral error caused or contributed to the crash." A 2008 NHTSA study using different data put the percentage at 93 percent.
5. National Safety Council. The 27 total percent for distracted driving is specifically for cell phone use, primarily talking or texting, according to NSC statistician Ken Kolosh. Other forms of distraction would push the percentage of accidents higher still, though data is scant. According to

NSC's annual report, *Injury Facts* (2015 edition), data shows that at any one time during daylight hours, 9 percent of drivers on the road are using cell phones while driving. About 10 percent of drivers admit to texting or e-mailing while driving.

6. *Unlicensed to Kill: Research Update*, AAA Foundation for Traffic Safety, March 5, 2008.
7. This is the case of Abby Sletten, whom prosecutors accused of looking at Facebook photos at the time of the crash instead of paying attention to the road. As often happens in such cases, the evidence could only at most show that the Facebook app on her phone was active, not that she was actually looking at it when she crashed.
8. "After 47-Second Hearing, Driver Who Ran Over 3-Year-Old Is Found 'Not Guilty'" by Andrea Bernstein, WNYC.org, November 19, 2014.
9. "Epidemic of Fatal Crashes" by Bill Sanderson, *Wall Street Journal*, February 10, 2014.
10. "Sober Drivers Rarely Prosecuted in Fatal Pedestrian Crashes in Oregon" by Aimee Green, OregonLive.com, November 15, 2011.
11. Prosecutors have no compunction about going after reckless fire starters or gun owners who negligently use firearms or leave them unattended for children to find. And neglectful parents routinely face a range of serious charges, from negligent homicide to child endangerment, as well as civil proceedings to supervise, limit, or sever their parental rights. The main exception: if the injury or endangerment occurs during a car crash, the perpetrator is unlikely to face any serious effort by the justice system to hold him or her responsible, or to deter similar behavior in the future. Cars and drivers, no matter how awful the circumstances, generally get a pass.
12. *Understanding the Distracted Brain*, National Safety Council White Paper, April 2012.
13. Ibid. The video can be viewed at http://www.mlive.com/news/grand-rapids/index.ssf/2010/12/video_grand_rapids-area_mom_of.html.
14. The National Safety Council, extrapolating distraction as a likely cause based on individual crash descriptions, puts the figure at 26 percent of fatal collisions. The National Highway Transportation Safety Administration, relying on official findings of distraction by police, attributes 10 percent of fatal crashes and 18 percent of injury crashes to distraction.
15. *Distraction and Teen Crashes: Even Worse Than We Thought*, AAA Foundation for Traffic Safety, March 25, 2015.
16. *Injury Facts*, National Safety Council.

17. "L.A.'s Bloody Hit-and-Run Epidemic" by Simone Wilson, *Los Angeles Weekly*, December 6, 2012.
18. "4 out of 5 Hit and Runs Go Unsolved in LA" by Jane Yamamoto, Tena Ezzeddine, and Keith Esparros, NBCLosAngeles.com, April 25, 2015.
19. "Fatal Hit-and-Run Crashes on Rise in U.S." by Larry Copeland, *USA Today*, November 10, 2013.
20. *Impact Speed and a Pedestrian's Risk of Severe Injury or Death*, AAA Foundation for Traffic Safety, September 2011.
21. Mass transit streetcars were similarly shunted aside as traffic impediments once a 1935 federal law forced electric utilities to divest their ownership of electric mass transit companies. This not only removed the streetcars' source of low-cost energy but also their political clout with the local politicians setting new traffic policies. The law drove hundreds of private streetcar companies out of business within a decade.
22. *An Empirical Analysis of Driver Perceptions of the Relationship Between Speed Limits and Safety* by Fred Mannering, Transportation Research, Part F (2008).
23. *NHTSA Finds Nearly Half of All Drivers Believe Speeding Is a Problem on U.S. Roads*, National Highway Traffic Safety Administration," December 11, 2013. The report also found that one out of five drivers admit, "I try to get where I am going as fast as I can."
24. The Vision Zero movement began in Sweden in 1997 as a rejection of the common practice of treating injury and loss of life like any other line item in a cost-benefit analysis of a traffic or road project. Vision Zero's core principle holds that life and health can never be bartered in exchange for a driver's convenience or a speedier parcel delivery. Instead, Vision Zero advocates argue that the overriding goal in road design should always be to preserve and enhance safety, health, and quality of life, with all other considerations secondary.

    Los Angeles's official explanation of Vision Zero is more pragmatic: "Vision Zero is a road safety policy that promotes smart behaviors and roadway design that anticipates mistakes so that collisions do not result in severe injury or death."

    In addition to Sweden, New York, and Los Angeles, other locations with Vision Zero programs include: The nation of Norway, and the following American cities: Chicago, Austin, Boston, Seattle, Portland, and the California cities of San Francisco, San Jose, San Mateo, San Diego, and Santa Barbara.
25. "2013 Motor Vehicle Crashes: Overview," *Traffic Safety Facts*, U.S. Department of Transportation, December 2014.

26. "Seat Belt Use in 2014—Overall Results," Traffic Safety Facts, U.S. Department of Transportation, February 2015.
27. "The 'Arms Race' on American Roads: The Effect of Sport Utility Vehicles and Pickup Trucks on Traffic Safety" by Michelle J. White, *Journal of Law and Economics*, University of California, San Diego, October 2004.
28. Pounds That Kill: The External Costs of Vehicle Weight by Michael L. Anderson and Maximilian Auffhammer, University of California, Berkeley.
29. KidsAndCars.org.
30. The Cameron Gulbransen Kids Transportation Safety Act of 2007, http://www.gpo.gov/fdsys/pkg/PLAW–110publ189/pdf/PLAW–110publ189.pdf.
31. This directive was issued one day before the U.S. Department of Transportation was due in court for trial in a lawsuit filed by the group Public Citizen, which sought to force the department to move forward on the 2007 law. Source: March 31, 2014 press release, "Government Finally Issues Rear Visibility Safety Rule for Vehicles, Will Save Lives After Years of Needless Delay," www.Citizen.org.
32. "Parents Share Story of Accidentally Backing over, Killing Son," KPTV.com, April 23, 2015.
33. Mothers Against Drunk Driving.
34. Title 49 of the United States Code, Chapter 301, Motor Vehicle Safety Standard No. 208—Occupant Crash Protection Passenger Cars, U.S. Department of Transportation, National Highway Traffic Safety Administration, January 1, 1968.
35. "Would You Buy a Car with a Built-in Breathalyzer?" by Kate Ashford, Forbes, March 25, 2015.
36. Uber.
37. CalTrans.
38. "MTA Bus Driver Charged After Running Over 15-Year-Old Girl in Brooklyn, Seriously Injuring Her Leg" by Pete Donohue and Barry Paddock, New York Daily News, February 14, 2015, and "Slap 'Em with Words: Anatomy of a Daily News Editorial on Vision Zero," Brooklyn Spoke.com, February 22, 2015.
39. "Transit Workers Ratify New Deal That Means Raises and Safety Measures but Higher Health-Care Premiums," New York Daily News, May 20, 2014.
40. The district attorney stated that charging the driver in this case with a crime under Oregon state law requires the driver to have been aware that

he was driving carelessly and nonetheless continued on, leading to the accident. Being oblivious leading up to the accident, it appears, constitutes an ironclad defense.

From the Lane County (Oregon) District Attorney's press release on the case:

"The Springfield Police Department and District Attorney's Office have concluded that Larry LaThorpe was traveling eastbound on Main Street, at or around the posted speed limit, when he unwittingly ran a red light at 54th Street. At that time, Cortney Hudson-Crawford was walking her three children in the west crosswalk southbound on Main Street. Tragically, by the time LaThorpe and Hudson-Crawford recognized the other's presence it was too late. Cortney Hudson-Crawford and her children were struck by LaThorpe's vehicle. LaThorpe remained on the scene and voluntarily complied with the investigation.

"Oregon courts have held that 'mere inadvertence, brief inattention, or errors in judgment' are not sufficient to charge a person with criminal homicide. In order for the State to prosecute a person for criminally negligent homicide a prosecutor must prove, at a minimum, 'that the defendant should have been aware of a problem with the defendant's driving, such as swerving, inattention, or near collisions, before the ultimate accident giving rise to the charges.' Because the investigation yielded no such evidence, criminal charges will not be filed against Larry LaThorpe for his role in the fatal crash on February 22, 2015."

41. "When a Tragic Accident Is Just a Tragic Accident," *Oregonian* Editorial Board, OregonLive.com, May 9, 2015.
42. "Just an Accident" by Charles Marohn, Strongtowns.org, June 9, 2015.
43. "$900 Million Penalty for G.M.'s Deadly Defect Leaves Many Cold," *New York Times*, September 17, 2015.

CHAPTER 6: PIZZA, PORTS, AND VALENTINES

1. Because of union rules on speaking with the media, Tom (not his real name) asked not to be named.
2. Domino's Pizza and "Domino's Pizza and Its Master Franchise Model" by Adam Jones, MarketRealist.com, March 26, 2015.
3. "Timeline: A History of GM's Ignition Switch Defect," *National Public Radio* (online), March 31, 2014.
4. "Dozens of Managers Were Involved in VW's Diesel Scandal," *Wired* (online), October 14, 2015.

5. This restriction is the FAA's long-standing rule for remote control aircraft, originally created for hobbyists, but now being applied to modern drone aircraft.
6. Sales totals for the 2014 model year, reported by automakers in 2015. Nationally, the top ten passenger vehicles according to Edmunds.com were:
    1. Ford F-150 pickup truck
    2. Chevy Silverado 1500 pickup truck
    3. Toyota Camry
    4. Honda Accord
    5. Toyota Corolla
    6. Nissan Altima
    7. Honda CRV
    8. Honda Civic
    9. Ford Escape
    10. Dodge Ram 1000 pickup truck

    In California, the top 10 were:
    1. Honda Accord
    2. Toyota Prius
    3. Honda Civic
    4. Toyota Camry
    5. Toyota Corolla
    6. Ford F-150 pickup
    7. Honda CRV
    8. Nissan Altima
    9. Nissan Sentra
    10. Silverado 1500 pickup
7. The Bakken Formation is a subterranenan layer of shale rock occupying 200,000 square miles beneath parts of North Dakota, Montana, Manitoba, and Saskatchewan. The oil-rich formation is about 400 million years old but the petroleum deposits were mostly unrecoverable before recent advances in hydraulic fracturing. By the end of 2013, Bakken oil production represented more than 10 percent of the U.S. supply. The formation draws its name from the farm on which it was first discovered; the owner was Henry Bakken of Tioga, North Dakota.
8. "Bakken Crude, Rolling Through Albany," *New York Times*, February 27, 2014.
9. A partial list of oil train crashes since mid-2013:
    July 5, 2013: A Montreal, Maine & Atlantic Railway train hauling Bak-

ken crude from North Dakota derailed, spilled 1.6 million gallons of oil, and exploded into flames in Lac-Megantic, Quebec. Forty-seven people died and thirty buildings burned.

November 8, 2013: A North Dakota oil train derailed and exploded near Aliceville, Alabama, spilling 749,000 gallons. There were no fatal injuries.

December 30, 2013: A Burlington Northern & Santa Fe oil train exploded near Casselton, North Dakota, forcing the evacuation of more than 2,000 from their homes and businesses.

January 20, 2014: Six CSX oil tank cars, part of a one-hundred-car train out of Chicago, derailed while crossing the Schuylkill River in a densely populated area of Philadelphia. The bridge is near the University of Pennsylvania campus and three hospitals, but no oil spilled or caught fire.

April 30, 2014: Fifteen cars of a crude oil train derailed and caught fire in Lynchburg, Virginia, spilling nearly 30,000 gallons of oil into the James River.

February 14, 2015: A one-hundred-car Canadian National Railway train hauling crude oil and petroleum by-products derailed in a remote part of Ontario, Canada. The blaze it ignited burned for days.

February 16, 2015: A 109-car CSX oil train derailed and caught fire near Mount Carbon, West Virginia, leaking oil into a Kanawha River tributary and burning a house to its foundation. The blaze burned for most of a week.

March 10, 2015: Twenty-one cars of a 105-car Burlington Northern & Santa Fe train hauling oil from the Bakken region of North Dakota derailed about three miles outside Galena, Illinois, a town of about 3,000 in the state's northwest corner.

March 7, 2015: A ninety-four-car Canadian National Railway crude oil train derailed about three miles outside the northern Ontario town of Gogama. The resulting fire destroyed a bridge. The accident was twenty-three miles from the February 14 derailment.

May 6, 2015: A 109-car Burlington Northern & Santa Fe crude oil train derailed near Heimdal, North Dakota. Six cars exploded into flames and an estimated 60,000 gallons of oil spilled.

July 16, 2015: More than 20 cars from a 108-car Burlington Northern & Santa Fe oil train derailed east of Culbertson, Montana, spilling an estimated 35,000 gallons of oil.

10. State of California.

11. Rockefeller Foundation Infrastructure Survey, 2011.
12. *40 Years of the US Interstate Highway System: An Analysis of the Best Investment a Nation Ever Made* by Wendell Cox and Jean Love, for the American Highway Users Alliance, June 1996.
13. "Bridging the Gap," *Economist*, June 26, 2014.
14. *The Global Competitiveness Report: 2014–2015*, World Economic Forum (online).
15. *Mapping Freight*, Brookings Institution.
16. American Society of Civil Engineers.
17. After years of withering criticism and fears of a deadly disaster should one of the tunnels collapse, New Jersey governor Chris Christie reversed course in the fall of 2015 and agreed to begin planning for what will be a ten-year tunnel replacement project. Under a tentative agreement, New York and New Jersey will bear half the cost, and federal funds will cover the other half.
18. Amtrak, America's long-distance passenger rail system, has an impressive-sounding 18,500 miles of track across thirty-nine states. But most of it is old, slow, and underused, incapable of accommodating the sorts of cross-country, high-speed, 200-mile-per-hour bullet trains running throughout Europe and Asia. Nearly a third of all rail passengers in the U.S. can be found on the 450 miles of the Northeast Corridor, home to America's only high-speed rail, the Acela Line, with a maximum speed of 150 miles per hour. That max is slow enough by international standards, but the corridor tracks are so poor that most of the Acela route drops below 100 miles per hour. Meanwhile China has eight hundred high-speed passenger trains crisscrossing the nation at twice that speed and covering more than twelve thousand miles—the world's largest bullet train network—while such countries as Spain, Japan, Germany, and France operate three hundred to six hundred bullet trains each. America has twenty.
19. Association of American Railroads.
20. "If the West Coast Ports Shut Down, Who Wins and Who Loses?" *Los Angeles Times,* April 13, 2015.

CHAPTER 7: THE LADIES OF LOGISTICS

1. From the 2016 fiscal year budget request for the U.S. Navy.
2. "The Top 100 League," Alphaliner.com, as of July 2015.
3. The shipping lines had originally planned a three-company alliance called P3 that also would have included the third-largest shipping line,

the French-based CMA-CGM Line with its fleet of 474 container vessels. The alliance would have controlled nearly 40 percent of the world's cargo carrying capacity. It had been approved by regulators in Europe and the United States, but China rejected the plan, expressing concerns about concentration of power in shipping and fears the "de facto merger" would stifle competition.

4. "Made in the USA: More Consumers Buying American," CNBC.com, March 6, 2013.
5. "Thirty-Nine Years of US Wood Furniture Importing: Sources and Products" by William G. Luppold and Matthew S. Bumgardner, *BioResources*, 2011.
6. *Review of Maritime Transport*, United Nations Conference on Trade and Development, 2014.
7. U.S. Bureau of Economic Analysis and "Hitting China's Wall" by Paul Krugman, *New York Times*, July 18, 2013.
8. Assuming the standard five-ounce can, using sales data reported in "US 2014 Shrimp, Canned Tuna Consumption Increases Seen Offsetting Declines in Other Species" by Matt Whittaker, UnderCurrentNews.com, January 30, 2015.
9. "The Alameda Corridor: A White Paper," *School of Policy, Planning and Development*, University of Southern California, June 2004.
10. New York Metropolitan Transportation Council.
11. American Trucking Association.
12. According to Operation Lifesaver, a rail-industry-sponsored organization culling data from the Federal Railroad Administration, there were 2,280 collisions between motor vehicles and trains at highway-rail crossings in 2014, killing 267 and injuring 832—the highest number of crashes and deaths since 2008. Still, the long-term trend has shown great improvement. There were 9,461 collisions, 728 deaths, and 3,292 injuries at crossings in 1981.

    Better safeguards led to this improved record, although only a third of the nation's 200,000 railroad crossings have flashing lights and moving barriers. Even those measures are an imperfect solution, as those crossings—which tend to be the busiest—still account for half of the car-train collisions every year. Drivers routinely go around the barriers or ignore warning lights as if they were merely inconveniences. Underpasses and overpasses that physically separate cars and trains are the only foolproof solution (other than closing a crossing entirely) but are usually deemed too costly or impractical.

According to Operation Lifesaver, a rail safety nonprofit founded by the rail industry:

- A typical 100-car freight train weighs the same as 4,000 automobiles combined.
- When a vehicle-train collision occurs, it is equivalent to a car crushing a soda can.
- Approximately every two hours, either a vehicle or a pedestrian is struck by a train in the U.S., an average of twelve incidents each day.
- More than half of all grade-crossing collisions occur where train speeds are 30 miles per hour or less.
- About two-thirds of all collisions at crossings in the United States happen in daylight.

13. Alameda Corridor Transportation Authority.
14. *Survey Results for Containers Lost at Sea—2014 Update*, World Shipping Council.
15. "EPA Bans Sooty Ship Fuel off U.S. Coasts," *Scientific American,* August 12, 2012.
16. "Prevention and Control of Ship and Port Air Emissions in China," Natural Resources Defense Council, October 2014.
17. European Commission, Climate Action.
18. Global Warming on the Road: The Climate Impact of America's Automobiles, Environmental Defense Fund, 2006. According to the report, global automobile emissions generate 10 percent of humanity's carbon footprint, with America responsible for almost half.
19. United Nations International Maritime Organization.
20. "Air Quality Report Card," published by the Ports of Los Angeles and Long Beach.
21. Compared to 2005, emissions at the Port of Los Angeles had dropped 87 percent for diesel particulate matter, 85 percent for particulate matter less than 2.5 microns, 87 percent for particulate matter less than 10 microns, 31 percent for nitrogen oxides, and 97 percent for sodium oxides. Source: Air Quality Report Card, Port of Los Angeles, 2014.
22. Time and cost averages were supplied by John Slangerup, director of the Port of Long Beach, based on March 2015 rates.
23. Knatz retired from the port in 2014 when the incoming mayor of Los Angeles sought to remake his predecessor's leadership team. Knatz continues to work on port issues as a University of Southern California professor, a member of the governance coordinating committee of the U.S. National Ocean Council, chair of the World Ports' Climate Initiative, and with

other professional organizations, as well as through her ongoing work with the Ladies of Logistics.

CHAPTER 8: ANGELS GATE

1. In later years the Marine Exchange became a nonprofit operation independent of the chamber of commerce.
2. The standard measure of a container ship's capacity is the TEU, the twenty-foot equivalent unit. So the *MSC Flavia* is listed as a 12,400-TEU vessel. There are a variety of container sizes in use around the world but by far the most commonly used—the standard, in practice if not in name—is the 40-footer. That container is the equivalent of 2 TEUs, so a vessel with a 12,400-TEU capacity can carry up to 6,200 standard containers.
3. According to the Tioga Group, as reported in "What's Needed to Ease Drayage Gridlock? Cooperation," *Journal of Commerce,* March 31, 2014.

CHAPTER 9: THE BALLET IN MOTION

1. Most major ports in the U.S. have transmodal freight yards somewhere close by, which unload the forty-foot standard shipboard containers into larger fifty-three-foot boxcar-sized containers, using the same sort of gantry cranes as the ports employ. This allows shorter, more efficient trains to be built for freight movement nationwide. There are several within a few miles of the Southern California ports, serviced by the drayage fleet.

CHAPTER 10: THE LAST MILE

1. UPS. This is a daily average.
2. For 2015. From "Essential Financial and Operating Information for the 100 Largest For-Hire Carriers in the U.S. and Canada," ttnews.com.
3. "At UPS, the Algorithm Is the Driver" by Steven Rosenbush and Laura Stevens, *Wall Street Journal,* February 16, 2015.

CHAPTER 11: PEAK TRAVEL

1. New Directions, 2013.
2. "Beyond Traffic," U.S. Department of Transportation.
3. "The 3-D Printing Revolution" by Richard D'Aveni, *Harvard Business Review,* May 2015.

CHAPTER 12: ROBOTS IN PARADISE

1. *Climbing Mount Next: The Effects of Autonomous Vehicles on Society* by David Levinson, Nexus Research Group, University of Minnesota.
2. "95 percent of All Trips Could Be Made in Electric Cars," GreenCar Reports.com, January 13, 2012.
3. "Parking in Mixed-Use U.S. Districts: Oversupplied No Matter How You Slice the Pie," Rachel R. Weinberger and Joshua Karlin-Resnick, Nelson/Nygaard Consulting Associates, August 1, 2014.
4. "Parking Infrastructure: A Constraint on or Opportunity for Urban Redevelopment?" Mikhail Chester, Andrew Fraser, Juan Matute, Carolyn Flower, and Ram Pendyala, *Journal of the American Planning Association*, 2015, 81(4).
5. "Driverless Cars and the Myths of Autonomy," David Mindell, *Huffington Post*, October 14, 2015.
6. "Heavy and Tractor-Trailer Truck Drivers," *Occupational Outlook Handbook*, U.S. Bureau of Labor Statistics, accessed online, November 2015.
7. "The Big Shift Last Time: From Horse Dung to Car Smog" by Andrew Nikiforuk, TheTyee.com, March 6, 2013.
8. "The Driverless Debate: Equal Percentages of Americans See Self-Driving Cars as the 'Wave of the Future' Yet Would Never Consider Purchasing One," Harris Poll, March 24, 2015.

   The poll found that "Americans are split on whether self-driving vehicles are safe for those inside them, but majorities see them as a danger to pedestrians and fellow drivers."

CHAPTER 13: THE NEXT DOOR

1. "Hit-and-Runs Take a Rising Toll on Cyclists," *Los Angeles Times*, November 29, 2014.
2. Hammer Conversations, Seleta Reynolds and Janette Sadik-Khan, March 26, 2015.
3. According to Janette Sadik-Khan.
4. *New York City's Red-Light Camera Program: Myths vs. Realities*, New York Department of Transportation, 2012.
5. Data from Houston Police Department, reprinted in "Crashes Doubled After Houston Banned Red Light Cameras," Streetblogs USA, August 12, 2015.

6. According to *Beyond Traffic: 2045*, published online by the U.S. Department of Transportation in 2015, even a relatively small 5 percent ridership on mass transit has an impact on traffic congestion: "By one estimate if public transportation services in the 15 largest metropolitan areas in America were eliminated and their riders shifted to private vehicle travel, it would result in a 24 percent increase in traffic congestion and cost our economy more than $17 billion annually."
7. *Who Drives to Work? Commuting by Automobile in the United States: 2013*, American Community Survey Report by Brian McKenzie, U.S. Census Bureau, August 2015.
8. "Frequently Asked HOV Questions," Federal Highway Administration.
9. *Modes Less Traveled—Bicycling and Walking to Work in the United States: 2008–2012*, American Community Survey Reports by Brian McKenzie, U.S. Census, May 2014.
10. "Pedometer-Measured Physical Activity and Health Behaviors in U.S. Adults," *Medicine & Science in Sports & Exercise*, October 2010.
11. "Physical Activity in an Old Order Amish Community," *Medicine & Science in Sports & Exercise*, January 2004.
12. *Mobility Plan 2035*, Los Angeles City Planning Department, May 28, 2015.
13. *The Decline of Walking and Bicycling*, National Center for Safe Routes to School.
14. "City of the Future," National League of Cities, Center for City Solutions and Applied Research, 2015.
15. Some critics complain that tolls, whether or not they are designed to discourage congestion, are a regressive tax that takes a greater proportion of the income of the poor than the rich (as opposed to a progressive tax, such as the U.S. income tax, which imposes higher rates on those with higher incomes). The critics are right about the regressive nature of tolls, but the entire system in use now—paying for roads through gasoline taxes, property taxes, registration fees, and, in many states, sales taxes—is already entirely regressive. A study that compared the impact of all these funding mechanisms for roads found that toll lanes were the least regressive tax because drivers still had the option to drive in adjacent free lanes: paying the tax was a choice, unlike the sales tax or the gas tax.

    The study's authors also argued that tolls were an inherently fairer method of paying for roads because they only targeted drivers. Sales and property taxes are less fair, the argument goes, because they impose the costs of driving on all taxpayers, even those who don't drive. This is a complicated and debatable position, however, as even non-drivers benefit from the goods movement and delivery of services that roads make possible. Source: "Just Road Pricing," by Lisa Schweitzer and Brian D. Taylor, *ACCESS*, Spring 2010.

16. "Too Big for the Road," *Governing*, July 2007; "The Hidden Trucking Industry Subsidy," *TrueCostBlog.com*, June 2, 2009; and "Overweight Trucks Damage Infrastructure," Associated Press (via *USA Today*), September 10, 2007.

APPENDIX

1. According to the nonprofit group PursuitSafety, police car chases lead to one death a day in the U.S., and about 150 of those deaths every year are innocent bystanders. The data on this is uncertain and tough to come by, as reporting of car chases gone wrong is strictly voluntary for police departments, and PursuitSafety asserts that the official tally kept by the National Highway Transportation Safety Administration is therefore incomplete.

   Even the incomplete statistics have generated concern and reforms. Many police agencies around the country have adopted restrictions on car chases in recent years in an attempt to reduce the risk of crashes, particularly for innocent bystanders, but also to avoid using dangerous, potentially lethal tactics for minor crimes. In years past, more than 90 percent of police chases were in response to nonviolent crimes—the sort of crimes for which police officers would not normally use deadly force. A 2008 data base analysis by the International Association of Chiefs of Police broke down the reasons for police chases by percentages:

   - Traffic violations started 42.3 percent of police car chases
   - Stolen vehicle: 18.2 percent
   - Driver believed to be intoxicated: 14.9 percent
   - Violent felony: 8.6 percent
   - Nonviolent felony: 7.5 percent
   - Other misdemeanors: 5.9 percent
   - Assisting other departments: 2.6 percent

   Thirty-five to 40 percent of these chases ended in a crash.

   Facts and Information About Vehicular Police Pursuits, PursuitSafety, March 2013.

Index

Index